Local Agri-food Systems in a Global World

Local Agri-food Systems in a Global World: Market, Social and Environmental Challenges

Edited by

Filippo Arfini, Maria Cecilia Mancini and Michele Donati

CAMBRIDGE SCHOLARS PUBLISHING

Local Agri-food Systems in a Global World:
Market, Social and Environmental Challenges,
Edited by Filippo Arfini, Maria Cecilia Mancini and Michele Donati

This book first published 2012

Cambridge Scholars Publishing

12 Back Chapman Street, Newcastle upon Tyne, NE6 2XX, UK

British Library Cataloguing in Publication Data
A catalogue record for this book is available from the British Library

ISBN (10): 1-4438-3664-8, ISBN (13): 978-1-4438-3664-7

CONTENTS

CONTRIBUTORS

Giovanni Belletti	Department of Economic Sciences, University of Firenze, Italy
Zora Bouamra-Mechemache	School of Economics, Toulouse, France
Jad Chaaban	Department of Agricultural Sciences, American University of Beirut, Lebanon
François Casabianca	Institut National de Recherche Agronomique, Corte, France
Marcelo Champredonde	Instituto Nacional de Tecnología Agropecuaria (INTA)/EEA Bordenave, Pigüé, Provincia de Buenos Aires, Argentina
Daniel Coq-Huelva	Department of Applied Economics II, Sevilla University, Spain
Sylvie Fanchette	IRD/CEPED, Paris, France
Christian Fischer	Faculty of Science and Technology, Free University of Bolzano, Italy
Xavier Gellynck	Department of Agricultural Economics, Ghent University, Belgium
Isabel Hervás-Fernández	Spanish National Research Council/Centre of Human and Social Sciences, Madrid, Spain
Virginie M. Lefebvre	Department of Agricultural Economics, Ghent University, Belgium

Allison Loconto Institut National de la Recherche Agronomique, Science et Société – Institut Francilien Recherche Innovation et Société (IFRIS) – Université Paris-Est Marne La Vallée – Champs sur Marne France

Maria Cecilia Mancini Department of Economics, University of Parma, Italy

Andrea Marescotti Department of Economic Sciences, University of Firenze, Italy

Evy Mettepenningen Department of Agricultural Economics, Ghent University, Belgium

Adrienn Molnár Department of Agricultural Economics, Ghent University, Belgium

José Muchnik Institut National de Recherche Agronomique-Syal ERG Innovation, Montpellier, France

Denis Requier-Desjardins Institut d'Etudes Politiques de Toulouse – LEREPS, Université de Toulouse, France

Florencio Sánchez-Escobar Regional Development Institute, Sevilla University, Spain

Javier Sanz-Cañada Spanish National Research Council/Centre of Human and Social Sciences, Madrid, Spain

Fabio Sforzi Department of Economics, University of Parma, Italy

Gerardo Torres Salcido Center for Interdisciplinary Research in Science and Humanities, Mexico City, Mexico

Valerie Vandermeulen Department of Agricultural Economics, Ghent University, Belgium

Guido Van Huylenbroeck Department of Agricultural Economics, Ghent University, Belgium

EDITORIAL ADVISORY BOARD

Preface

The idea for this volume was conceived after the International EAAE-SYAL Seminar "Spatial dynamics in agro-food systems: implications for sustainability and consumer welfare" held in Parma, Italy, in October 2010. The Seminar was the result of the joint efforts of two scientific communities: SYAL—ERG (Systèmes Agro-alimentaires Localisés—European Research Group) and the European Association of Agricultural Economists (EAAE), closely involved in the debate on the dynamics of local production systems under the pressures of globalized food-chains with respect to environmental preservation, cultural identity, market equilibrium and in general socio-economic welfare.

The awareness that an effective analysis on the spatial dynamics within the agri-food systems requires an interdisciplinary approach led to the involvement of scientists from the fields of economics, geography, sociology, demographics and agronomy. These fields can all provide useful theoretical tools and innovative analytical outcomes.

We would like to express our gratitude to the International Scientific Committee of the EAAE-SYAL Seminar, in particular to José Muchnik, Javier Sanz Cañada, Corrado Giacomini and Gilles Allaire, who helped to make it possible to explore the complexity of the local agri-food systems in the present era.

Although there is a great deal of literature on local agri-food systems, the links between local and global strategies of food production and processing are in rapid evolution and we felt an urgent need to update the underlying theoretical debate. We therefore gathered together contributions covering the following areas: the current state of spatial dynamics in agri-food systems; the socio-environmental impact of agri-food systems on rural development; the role of local resources in agri-food systems; the governance and public policies of local agri-food systems; and, last but not least, the progress in methodological approaches for analyzing spatial dynamics of agri-food systems.

The aim of this volume is to further interdisciplinary discussion, and it is hoped, to propose new pathways for the development of sustainable local agri-food systems in today's globalized world.

—Filippo Arfini, Maria Cecilia Mancini and Michele Donati

INTRODUCTION

LOCAL AGRI-FOOD SYSTEMS IN A GLOBAL WORLD: MARKET, SOCIAL AND ENVIRONMENTAL CHALLENGES

FILIPPO ARFINI, MARIA CECILIA MANCINI AND MICHELE DONATI

The purpose of this volume is to present a broad-based critical analysis of the relationship between the agri-food sector and the socio-economic environment when local agri-food production systems are connected with global markets. The key role of this topic was confirmed by a great deal of literature and it has been widely discussed as part of the complexity of agri-food systems, involving close connections between different dimensions, institutions and frameworks.

It is a fact nowadays that food production, food processing and food consumption are economic activities in which local and global strategies are interconnected (Goodman 2004; Bowen 2011) and they are combined in complicated ways (Murdoch et al. 2000; Sonnino 2007). Decisions on where and how to produce, associated with what to consume, are made on a global scale. They are driven by the cost differentials of factors such as labour and transportation, but also by target market characteristics including quality needs, beliefs and cultural heritage.

In agri-food as in other systems, many companies have organized production in developing countries in order to be more competitive on the global market. This process generates positive and negative externalities which can affect the welfare of the local population and, more generally, the sustainability of local production systems.

On the other hand, local production systems are competing on the global market by producing specific quality goods, where innovation, services, culture and local heritage are part of the concept of quality. They also affect the welfare of consumers and citizens; if products fail on the market, perhaps because of exogenous conditions, feed-back mechanisms will affect local production systems.

In a globalized word, it is possible to locate production activities in a limited geographical area and reach markets that are global (Murdoch and Miele 1999). These may be commodity markets, but in other cases they are niche and spatially concentrated markets. As a result, there is a wide range of agri-food systems showing a striking coexistence of diverse organizational patterns, ranging from simple food chains to more complex food networks. Agri-food systems always however develop according to the cultural and productive characteristics of local production systems as well as the availability of resources like land and water, production costs and the localization of target markets.

The implications of the relationships of different agri-food systems on local production systems are wide and varying. They fall into different areas: geography, economics, demography, sociology and agronomy which are all fields useful in assessing the impact of different behaviors and strategies on the socio-economic evolution of local production systems.

In this book, the local dimension of agri-food systems is considered assuming that all the variables related to agri-food production are closely connected and dependent on local production systems. In this perspective, Local Agri-Food Systems (LAFS) also become the focus of the analysis in order to evaluate the level of sustainability and policy requirements. Whether LAFS produce, transform or simply consume food, there are big implications for the socio-economic environment and socio-economic sustainability.

Although there is a great deal of literature on LAFS, there are still unresolved issues concerning the factors influencing their dynamics and effects, at both local and global level. The precise questions asked reflect different research objectives and policy goals, but the key issues are the following:

- what are the dynamics of LAFS under the pressures of globalized food-chains with respect to environmental preservation, cultural identity, market equilibrium, firm competition and, in general, socio-economic welfare?
- are LAFS a new paradigm for rural development?
- what is the role of local resources in local agri-food systems?

- what is the role of institutions and public policies in supporting LAFS?

The main issue for researchers however remains the definition of the theoretical framework able to catch most of the existing relationships between local and global, and provide answers on the above socio-economic phenomena.

The volume examines the interrelation between LAFS and agri-socio-economic aspects from two main points of view. It therefore consists of two parts: theoretical analysis of the different models at the basis of LAFS and empirical evidence of the role played by LAFS on local and global markets.

The theoretical debate on local approaches in the globalisation era

The theoretical debate in this book underpins many of the different types of analysis on the relationships between local and global. Mardsen and Murdoch (2006) pointed out the complexity of interaction and the presence of heated debate between the actor-network approach and another approach closer to political economics. The debate is in fact still polarized around these two approaches, and this is reflected in the different theoretical frameworks used by our contributors.

Sforzi and Mancini examine industrial district theory. This theory is well known, and is used to interpret industrial organization models developed at a local level. But industrial theory concepts are not widely applied to analyzing local food production systems. Sforzi and Mancini consider the theory as a possible framework for LAFS analysis. The focus of the analysis is the local community, but the key role is played by human agents of production and their knowledge. The agri-food system can be seen as "a global network of places", where each place is specialized in a different component of the system and where policy action meets the needs of local population in regard to global pressure.

Fisher on the other hand focuses on global value chains and the organizational relationships that effect sustainability at local level. It is a fact that the global value chain is the strategy developed by food chain managers in order to consider demand and supply equally with the objective of maximizing consumer company utility. In this respect, local producers of non-regional foods might suffer. On the other hand, effective food chains can also guarantee the transmission of information along the chain. The effectiveness of global value chains as a tool able to guarantee fair conditions to producers is based on the ability of food managers to develop an effective communication strategy involving local producers and other organizations. If they are able to do this, all members of the food chain raise profits and sustainability becomes a realizable goal.

Mettepenningen, Vandermeulen and Huylenbroeck analyse the theme of rural economics and rural development showing positive and negative externalities in today's globalized context. They consider regional identity and local food systems as able to boost rural development. The connections between these two relatively new elements are regional specificities, including the skills to provide an identity to local food, and the improvement of environmental quality. Here too there is the need to satisfy the expectations of consumers and citizens. This approach is somewhat similar to industrial district theory as the condition for this type of development process in rural areas involves the different stakeholders and policy support, the decentralisation of governance structure and a big investment in human resources at local level. The main question is who is responsible for institutional mechanisms able to influence the supply of private and public goods and avoid market failure.

Belletti, Casabianca and Marescotti indirectly support the approach of Mettepenningen, Vandermeulen and Huylenbroeck, as they consider the role of local food quality and local resources the key element in a sustainable approach for LAFS aiming to meet societal requirements for positive environmental externalities in production.

Torres-Salcido and Muchnik highlight the presence of paradoxes in the very complex behavior and strategies adopted by both consumers and private/public stakeholders that affect LAFS in a globalized framework. From the consumer point of view, the paradox is the fragmentation of consumption, where consumers are increasingly attracted by products with specific characteristics associated with a territory and their intrinsic value. From the stakeholder point of view, the paradox is the combination of collective action and social exclusion. The combination is leading to new inequalities and asymmetries in the organization and collective action inside and outside local communities, and leading to economic and social fragmentation. This criticism is closely related to the comments of the previous contributors and underlines the need for public policy to prevent failure on global markets and at the same time prevent social exclusion at the local level.

Lefebvre, Molnár and Gellynck conclude the first part of the book with a review of the literature on network performance and the factors influencing it. They show that inter-organizational networks are generally investigated for their impact on the performance of the firms participating in the networks. They also find that that where goals are considered, these generally refer to the goals of the firms participating in the network and not to the goals of the network itself. They ask whether inter-organizational goals really exist.

There is a clear link between different approaches and the need to create value for companies, consumers, society and the local community and, at the same time, there is a need to create and to manage a sustainable agri-food food system where development is not only based on economic values, but also on human and environmental value. The complexity and the problems of LAFS in a globalized framework are all up for discussion. The breadth and scope of the contributions to Part I show that there is no single framework able to embed all the factors affecting LAFS sustainability and responding at the same time to the needs of the global market.

Evidence of local complexities and new-socio economic patterns

Part II contains six chapters placing the issues theorized in the Part I into a global context. Each chapter presents a case or experience exemplifying the complexity and links between different factors influencing LAFS sustainability and their evolution over the time.

Some contributions show how LAFS take different forms and follow different dynamics in relation to local resources, market characteristics and policy strategies. The cases are from far-flung areas of the globe - Latin America (Desjardins), Spain (Sanz Cañada, Sánchez Escobar, Hervás Fernández and Coq Huelva), Tanzania (Loconto), Argentina (Champredonde and Muchnik), France (Mechemache) and Vietnam (Fanchette) and it is clear how, in a globalized context, geographical position affect LAFS characteristics, but it is also clear that local resources, market strategies and policy instruments used by different types of stakeholders are even more relevant. The cases demonstrate how socio-economic sustainability and environmental sustainability are closely linked. They are the result of how the different dimensions of the LASF combine with regard to objectives of local stakeholders.

The six cases illustrate different aspects of the behavior of LAFS and their level of sustainability:

- the adoption of a new paradigms of rural development (Desjardins);
- the adoption of R&D strategies to enhance environmental quality Sanz Cañada, Sánchez Escobar, Hervás Fernández and Coq Huelva);

- the impact of certification schemes on the governance of global value chains (Loconto);
- the dimension of quality for regional food products (Champredonde and Muchnik);
- the relationships between agri-food intervention mechanism, such as quality labels, and rural development (Mechemache);
- the adoption of innovation pattern and local policy strategies for global scale competition (Fanchette).

The contributions show that no matter where the LAFS is located, globalization brings with it common problems. In this there is little difference between the north and south of the world or between industrialized and under-developed countries. LAFS all over the globe present problems in common, which by implication may have common solutions too. The differences are due to the socioeconomic context, environmental factors, local heritage and the culture of the population, the skill of the entrepreneurs and the role and the power of public and private stakeholders.

This volume discusses the possibility that all over the world LAFS could be set on a common course towards development and sustainability.

References

Bowen, S. (2011). The Importance of Place: Re-territorialising Embeddedness. *Sociologia Ruralis* 51 (4): 325-348.

Goodman, D. (2004). Rural Europe Redux? Reflections on Alternative Agro-Food Networks and Paradigm Change. *Sociologia Ruralis* 44: 3-16.

Marsden, T. and Murdoch, J. (2006). *Between the local and the global. Confronting complexity in the contemporary agri-food sector*. Vol. 12. Elsevier: Amsterdam.

Murdoch, J., Marsden, T. and Banks, J. (2000). Quality, nature, and embeddedness: some theoretical considerations in the context of the food sector. *Economic Geography* 76 (2): 107-125.

Murdoch, J. and Miele, M. (1999). "Back to nature": changing "worlds of production" in the food sector. *Sociologia ruralis* 39(4): 465-484.

Sonnino, R. (2007). Embeddedness in action: saffron and the making of the local in southern Tuscany. *Agriculture and Human Values* 24 (1): 61-74.

PART I

CHAPTER ONE

A REINTERPRETATION OF THE AGRI-FOOD SYSTEM AND ITS SPATIAL DYNAMICS THROUGH THE INDUSTRIAL DISTRICT

FABIO SFORZI AND MARIA CECILIA MANCINI

Abstract

Industrial district theory has brought to development economics the opportunity to interpret economic change through places where it actually takes place, as a result of join action of local and extra-local social, economic and institutional forces. This chapter sets out to discuss the contribution that industrial district theory can make to the debate on spatial dynamics of agri-food systems in the age of globalisation. The first part analyses the contribution of the industrial district approach in the relationship between industry and territory; the second part studies the evolution of the concept of agri-food system and the main determinants of the spatial dynamics in modern agri-food systems. The authors agree that the industrial district theory can shed new light on the spatial dynamics of agri-food systems, and can offer an alternative to the mainstream approach. In using the local community as a unit of analysis, industrial district theory gives a key role to human agents of production and their knowledge and the agri-food system can be seen as "a global network of places", each place being specialized in a different component of the system.

1. Introduction: the industrial district and its various meanings

The industrial district (ID) is widely recognized as a *model of production* by the international academic community in that it reveals the capacity of small and medium enterprises (SMEs) to attain the same level of competitiveness as large firms. It represents a model of economic growth and social development.

Since the 1990s, regional authorities have used IDs as an *instrument of industrial policy* on the initiative of central government, which has promoted their legal recognition. In the first decade of this century, the ID was adopted as a theoretical background for the policy of *agrupaciones de empresas innovadoras* (Innovative Business Groupings) in Spain (MITYC 2006). In developing countries, the Italian experience of IDs has been used as a reference model by UNIDO (2001) for policies encouraging cooperation among artisan firms. Currently ID is the object of study and monitoring in many regions of Italy by academics and professionals working for public and private institutions.

The ID concept has also informed similar concepts such as the *systèmes productifs localisés* (localised production systems) (Courlet 2008), and to some extent introduced a "territorial dimension" into concepts such as the *cluster* (Porter 1990; Porter and Ketels 2009). The intellectual debt of these concepts to ID is widely recognized in the literature.

ID theory has also breathed new life into industrial and regional economics. Most recently it has brought to development economics the opportunity to interpret economic change through places where it actually takes place, as a result of combined action by local and extra-local social, economic and institutional forces.

Because of these multiple meanings—as the French economist Claude Courlet (2006, 20) noted—the original ID concept has "lost scientific rigor" or has been misinterpreted.

As a matter of fact, defining an ID as a geographical concentration of industry is equivalent to defining it through the theoretical framework of the location of industries. But that is precisely the theoretical framework called into question by ID theory; an ID comprises a local community specialized in an industry instead of an industry concentrated in a place.

Here is an example to clarify this point. Langhirano is a place well known to gourmets because it produces "Parma Ham". Langhirano can be seen either as one of the places where the food industry is localised, or as a mountain community near Parma which procures what it cannot produce itself by specializing in what it makes best. In the first view, the unit of analysis is the food industry, and a study of its spatial distribution reveals Langhirano. In the second view, the unit of analysis is the community of Langhirano, and research on the production structure of the place brings into focus the food industry. In the first view, the socio-economic reality is seen as "an array of interrelated industries" and in the second view it is "a system of places".

International historiography (e.g., Daumas 2007) is unanimous in recognizing that the ID concept began to spread among the scientific community thanks to the 1979 article by Giacomo Becattini, the Florentine economist founder of the Italian neo-Marshallian school.

Becattini's 1979 article was entitled "From industrial 'sector' to industrial 'district'. Some remarks on the unit of analysis of industrial economics" (now reprinted in Becattini 2004) are of interest. The title leaves no room for doubt; Becattini proposes the *district* in the place of the *sector* as a unit of analysis.

It is true that the article discusses economies of production and how

> the advantages of production on a large scale can in general be as well attained by the aggregation of a large number of small masters into one district as by the erection of a few large works (Marshall, cit. in Becattini 1979).

This clearly paves the way for the ID to be conceptualized as a model of production. But the central nucleus of ID theory remains that the ID is the unit of analysis of industrial phenomenology.

This chapter sets out to discuss the contribution that industrial district theory can make to the debate on spatial dynamics of agri-food systems in the age of globalisation.

Section 2 explains the theoretical foundations for this work. It examines the contribution of the ID approach in the relationship between industry and territory. It investigates the increasing importance given to territory in the analysis of the agri-food systems, highlighted by the growing number of theoretical approaches that use the territory as the explanatory variable of the socio-economic dynamics in modern agri-food systems, such as the *systèmes agro-alimentaires localisés* (hereafter: SYAL). Section 3 tackles the issue of spatial dynamics in agri-food systems and their determinants. Section 4 discusses a reinterpretation of the agri-food system and its spatial dynamics through ID theory and Section 5 proposes some final remarks.

2. A theoretical approach to the relationship between industry and territory

2.1 The ID as theoretical framework

We first need to consider the theoretical importance of taking the ID as a unit of analysis. ID theory breaks with mainstream economics, which sees the relationship between industry and territory as the result of industrial location. ID theory introduces territory into economic analysis, and overturns this perspective.

In contrast with studying the distribution of industries over a territory, searching for the presence of manufacturing agglomerations and related locational factors, ID theory takes as starting point the place where the economic agents (both entrepreneurs and workers) are embedded.

The approach is justified by the way in which industry is conceptualized. Instead of defining industry according to the traditional technological criterion, that is as "the set of firms which produce a given class of goods sharing the same technical characteristic or material of the manufacturing process", we conceptualize industry as "the awareness that economic agents have of belonging to an industry", through a criterion of identity.

Giacomo Becattini, in his 1962 book on the concept of industry, writes:

> As well as a set of manufacturing processes, industry may be seen as a set of sacrifices of work, expectations and organisation (Becattini 1962, 23).

In this view, the place of the industry is not a *production system*, that is a territorial partition circumscribing a group of firms in the same or related industry, but has the nature of a *place of living*, that is a bounded territory where a group of people live and earn their living in economic activities located in it. Most daily social relationships occur, and the need for social integration is met, within this same place.

The economic agents' awareness of belonging to an industry is formed inside the place (or local community) through the way in which the networks of local relationships (internal markets) intermingle with the networks of connections established with their customers and suppliers which may be national or even global (external markets). This system of internal/external networks between economic agents is also influenced by relationships with family and institutions, and their links with social and institutional contexts. All these aspects lead to the sharing of representations, norms, values and sanctions which provide a framework and background for the social and economic life of the local community.

It is this local community, relatively self-contained economically, socially and institutionally, which is the unit of analysis and which we label "industrial district".

2.2 The role of "territory" in the agri-food system

The relationship between agri-food system and territory has been gradually defined over time. Initially, the agri-food system was interpreted through the notion of *agribusiness* in order to bring out the role of farming and its upstream and downstream activities in the industrial processing of food products. Davis and Goldberg (1957) for example examined only aspects of production and in this way gave farming a key role in advanced capitalist economies. At the beginning of the 1970s, the French school studied the industrialization of farming and focused on its role in fulfilling the requirement for food (Malassis and Bourdon 1970). It was found that the consumer drove the engine of the agri-food business, and the attention of agrarian economists shifted from supply to demand.

In Italy the work of the French school was further developed by Galizzi (1975), who defined the field of agri-food study as "the set of functions which jointly satisfy a new requirement for food". Galizzi correctly predicted that supermarkets would gain precedence in the relationship with the consumer and dominance over the other components of the agri-food system.

Many subsequent researchers were to agree that the primary function of the sector is "feeding". Ghersi and Bencharif (1995) for example write:

> The agri-food system is made up of a set of agents in dynamic interaction who act on the production and transfer of food products in order to ensure food supply.

Ghersi and Bencharif thus recognise the complexity of the relationships between different, evolving, components of the agri-food system: farming, processing firms, suppliers of goods and services, food distribution, catering, consumers and institutions governing the agri-food system.

Again thanks to the contributions of the French school, territory was recognized in the 1990s as playing a key role in the agri-food system. As a consequence of this realization, the natural environment (i.e., the use and preservation of natural resources) and the socio-cultural environment (i.e., local history and manufacturing traditions), were also given new consideration. Both types of environment affect the production system and the tangible and intangible quality of a product. Food production was now interpreted as the result of a production model where economic agents are linked through the characteristics of a specific territory. Cultural and social aspects, and the collective dimension of production know-how of foods gave rise to a very rich field of study (Sylvander and Lassaut 1994; Berard and Marchenay 1995; Letablier and Delfosse 1995; Sylvander 1996; Casabianca and Valceschini 1996; De Sainte Maire 1996; Berard and Marchenay 1997).

The influence of the territory becomes particularly relevant in specialty foods. There are three types of factors involved: the specifically local nature of resources, the history and traditions, and the collective dimension of knowledge shared locally (De Sainte Maire et al. 1995; Sylvander 1995; Bérard and Merchenay 1995; Barjolle et al. 1998; Casabianca et al. 2005).

The relationship between agri-food system and territory allows us to separate the agri-food system into constituent subsystems; product, consumption, institutions and territory (Bertazzoli et al. 2006). For each subsystem the characteristics need to be specified. These include the type of firm, characteristics of products, variety of services, type of institutions etc. It is also necessary to specify the function (productive, cultural, social, political, landscape) as well as the relationship between these aspects. The consequence is that there is not just one single agri-food system. There are many different systems, and each one is defined and reproduced according to characteristics and links among the subsystems.

2.3 From the agri-food system to localised agri-food systems

The use of the variable "territory" for the study of local production initiatives led to the proliferation of classifications.[1]

A relevant approach to interpreting the relationship between the agri-food system and the territory is the French *systèmes agro-alimentaires localisés* (SYAL). This definition incorporates the territorial dimension of the system and, like previous definitions, allows for a plurality of situations. The concept was first formulated in the mid-1990s by the CIRAD (1996) and since then has gradually been refined (Muchnik 2010). There is no space here to summarise the evolution of the concept, but it is important to note that the concept of SYAL is related to another concept formulated in France, the *système productif localisé* (SPL) by Courlet and Pecqueur (1992), and developed and consolidated by Courlet over the following decade (Courlet 2008). That there are certain similarities between SYAL and SPL is confirmed by a recent comparative analysis by Requier-Desjardins (2007). Requier-Desjardins shows, though, that SYAL and SPL involve a different definition of the relationship between economic activity and territory. For a SPL the industry has to be concentrated in a relatively small area, a single place, while for SYAL

> the notion of geographical concentration given the dispersion typical of rural areas must be softened: spatial limits of SYAL may be quite wide, embracing sometimes an entire region, or a set of micro-basins in a region, a kind of archipelago (Requier-Desjardins 2007, 11).

Here are a couple of examples from the Parma area. The pig haunches to be turned into *Prosciutto di Parma* come to Langhirano from pig farms located in various regions of Italy, so the SYAL is a very wide area. But for producing *Parmigiano-Reggiano*, the milk has to come from the same area that the cheese-making, ripening and sale take place, so the SYAL is geographically much smaller.

An industrial economist could make a similar point about Biella wool. Biella is a local community located in North-east Italy, but the raw material does not actually come from the Biella area; it comes from sheep farmed in Australia. Biella is still however a local system specialized in woollen textiles even if some of its production stages are located outside the local system. This is because an industrial economist makes a distinction between the *industrial district* of Biella and the *economic space* of the industrial district of Biella, which is defined by the networks of trade with suppliers and even with final consumers located outside the district.

So the agri-food system is not unusual; its features are shared by production systems of other manufacturing sectors. In this view, a *localised agri-food system* coincides with its *economic space*, so its components (producers, manufacturers, retailers, consumers) may belong to any territory. The SYAL has a multi-localised nature with regard to territory, which is typical of a sectorial approach. But the territory does not define the production system, it only describes it.

3. Determinants of spatial dynamics in agri-food systems

The spatial dynamics characterizing modern agri-food systems are mainly the result of social, economic, cultural, technological, and institutional change. The way in which agri-food systems reorganise to manage change underlies their spatial dynamics, and is a cause rather than an effect of the current globalisation.

Hirst and Thompson (2003, 17) write:

> Globalisation has a history. The 50 years between 1950-2000 are not remarkable when compared with the period 1850-1914. In that period flows of merchandise trade, capital investment and labour migration were all comparable to or greater than those of today.

But it is undeniable that one of the main features of contemporary globalisation is the ease with which production processes can be divided into stages, locating each stage in places throughout the world according to cost advantage, and implementing remote monitoring of production through wireless technology.

The main changes in agri-food systems are occurring in four areas:

– new models of consumption;
– modern retail;
– technical progress and information;
– international regulations.

3.1 New models of consumption

Changes in the economy, society and demographics have led to the requirement for food products with high service content.

The increasing participation of women on the labour market, and the fact that women are now not entirely responsible for producing meals in the home means that there is today increased demand for time-saving products. At the same time, changes in labour organisation such as greater distances travelled between home and work and the decline of longer lunch breaks are making meals less important and increasing demand for snack products. Increased consumption of food outside the home and the growth of catering chains, immigration, population ageing and an increase in the number of single person households and single parent families are all factors which have led to a wider range of agri-food products being offered and stimulated the development of new sales formats and packaging.

A second type of variables affecting consumers is cultural. Food consumption today is no longer simply a question of nutrition; it is also an expression of lifestyle and personal values. The desire for physical well-being has led to increased demand for "light" products, fresh rather than processed, fruit and vegetable rather than meat-based, GM-free, organic foods and "novel foods" enriched with nutrients. New awareness of the environment is encouraging the consumer to choose low environmental impact products, organically and sustainably grown foods, and to save food miles by using local distribution channels and the short supply chain. Ethical considerations are making fair trade models more widespread in trade with poorer countries.

Consumer choice is thus influenced by a range of socio-economic demographic and cultural variables acting jointly on decision-making processes. Demand for local speciality products, for example, is the result of consumer sensitivity to the advantages of buying from local producers as well as appreciation of chemical-physical and taste-smell characteristics of products. Another example of converging requirements is the spread of fast food and catering chains which satisfy the requirement to save time and at the same time the desire to imitate consumption styles imported from abroad. And naturally, the price variable remains a basic criterion for the majority of consumers who are obliged to behave rationally in spending their income (Belletti and Marescotti 1996).

3.2 Modern retail

In the agri-food supply chain a downstream shift is currently taking place in the formation of value; the composition of agri-food products is increasingly influenced by those links in the chain closest to the final consumer. The most important of these actors in recent years is modern retail which has consolidated a relationship of loyalty with the consumer and replaced the traditional role of the manufacturer.

The use of private label in particular has enabled modern retail to intercept the relationship of the consumer with the manufacturer, by reinterpreting consumer requirements and providing a personal guarantee of products on offer. This strategy has revolutionised the supply chain relationship between manufacturer and retailer. It has become hard for the consumer to correctly identify the producer, who can thus be easily replaced by the retailer investing the image of the supermarket or chain in becoming owner of a brand name.

Whereas traditionally the manufacturer was considered to be the main operator, and the wholesaler or the retailer the agent,[2] the new relationship sees the supermarket in the main role. The supermarket now has vertical control of the chain; it controls to a large extent pricing policy upstream for the agri-food manufacturers and farmers, and defines the characteristics of the product it sells. Moreover, the supermarket often controls supplier output through “supply chain contracts” and/or imposing its own standards (Benoun and Héliès-Hassid 2003; Giacomini et al. 2010). These standards concern health and hygiene requirements, quality guarantees and environmental and animal welfare requisites, and agreeing to them is often a necessary condition for suppliers wanting to sell to modern retail (Arfini and Mancini 2004; Fulponi 2006; Fulponi et al. 2006; Henson and Reardon 2005; Giraud-Heraud et al. 2006; Valceschini et al. 2005). Many were set up by large retailers in Europe and have now spread all over the world. For example, the GlobalGAP standard has now been adopted in countries of Latin America, Africa, Asia and Oceania in response to the growing demand for out-of-season fruits and vegetables on western markets.

Such private certification schemes are useful to modern retail in two ways. As well as giving them vertical control of the supply chain, they ensure that retailers can offer differentiated range of products (Hatanaka et al. 1996), available on the market for both mass and niche consumption models.

The appearance of alternative consumer goods such as green foods, health foods and local specialty foods in supermarkets implies that modern retail standards today are used both in alternative and conventional supply chains in developed and developing countries.

The vertical control of modern retail chains over production constitutes a new form of agri-food governance which is overtaking public regulation on quality and food safety with private forms of control and guarantee (Farina et al. 2005; Bush and Bain 2004) and strengthening the oligopoly of supermarkets (Campbell et al. 2006).

3.3 Technical progress and information

Agri-food chain actors are today using new models of organisation to remain competitive in their response to consumer preference. A key role in competitiveness is played by technical progress applied to production and preservation of foods as well distribution and telecommunication. This process of "technicalisation" of food production, together with the recent series of food scares and the important presence of credence attributes in food products[3] has led to a degree of breakdown of trust among consumers. It is now commonplace for consumers, in a sort of self-defence, to search out information on credence attributes as well as the origins and nutritional value of a food product. So the dissemination of information has come to play a key role in regaining and strengthening consumer trust of producers and institutions. This is why there is now a proliferation of communication regarding food: there are specialized magazines published by consumer associations or supermarkets themselves, product certification schemes which are sometimes voluntary and public rules and regulations.

3.4 International regulations

Trade policies, food product quality standards and the environmental sustainability of production processes are all increasingly subject to regulation, while the role of the state in promoting this legislation is being increasingly supplanted by international organisations. There are several facets to international legislation. Its intention is to ensure food safety for consumers but at the same time it affects the organisation of companies and the food chain, and in the final analysis the balance of power and profit distribution along the supply chain (Hammoudi et al. 2009).

Busch and Bain (2004) find that the intervention of the WTO, which has brought an intense activity of regulatory standardization along with the liberalization of international trade, has contributed to the proliferation of private standards. Imposed by the large-scale retail organisations in order to guarantee product compliance with the growing body of legislation, private standards have become a tool for the large-scale retail organisations in strengthening their role in agri-food systems.

In the same way, EU food safety regulations have also led to a change in the composition of the agri-food chain. EU regulations have favored big firms at the expense of smaller ones, as these have not been able to exploit large-scale economies in complying with new regulations, and have thus become less competitive (Loader and Hobbs 1999).

3.5 Spatial dynamics in agri-food systems

The heterogeneous nature of consumer requirements is reflected in the supply of a wide range of agri-food products for which production stages take place overseas, even though production and distribution chains may be organised differently.

Products with a high service content are usually the output of an agri-food chain where crop cultivation or farming and initial phases of processing take place in the country of origin, often a developing country. On the other hand, industrial processing, which adds the most value, usually takes place near the final market in developed countries.

Low production costs, the complementary seasonal availability of products, speed of freight and the increasing capacity of developing countries to produce in the quality and quantity required by large-scale retail organisations are all factors making the spatial separation of the different phases of production across different countries more frequent. In the words of Grossman and Rossi-Hansberg (2006; 2008), networks of *trade-in-tasks* are replacing networks of *trade-in-goods*.

And this process of separating components of the agri-food chain on the global scale is one of the main manifestations of the internationalisation of economic activity.

4. A reinterpretation of the agri-food system and its spatial dynamics through ID theory

Nomen omen (true to its name), says the Latin expression, and the agri-food system is just that. Whether it is defined as localised, or stands without an adjective, the agri-food system has a sectorial focus. It is a production system which is geographically spread, not a system of places that, at a given time, are specialized in one or a few parts of the same process of production, processing, distribution and consumption.

In the traditional analytical framework, the territory is (a) a sub-system, when used to indicate a component of the agri-food system, in the same way as firms, the product, consumption and institutions; (b) a secondary category of analysis when used to indicate the origin of a product or the location of production stages.

In the framework of ID theory, the territory is (a) a local community in which economic agents are embedded; (b) a factor which modifies productivity and innovativeness deriving from the way the local community relates to the apparatus of production supplying "the social climate and the human factor".

However, research on a place (or local community) does not focus merely on internal factors such as inter-firm cooperation, the way that local population supply the apparatus of production with new entrepreneurs and employees and institutional provision of public goods. It also investigates external relationships with other places: those which are the sources of raw materials or semi-processed goods (place of production) and those which are final destinations of finished goods (place of consumption).

So, reinterpreted through ID theory, the agri-food system can be seen as "a global network of places", each place being specialized in a different component of the system. The spatial dynamics of the agri-food system, or the organisation of the network, depends on the social, economic and institutional dynamics of each place comprising the network. The organisation varies according to the number of involved places and consequent inter-local co-operative relationships, and according to the hierarchical order among them.

It is important to note that the impulse for change can appear at any point on the network, in production, processing or consumption. Section 3.1 above described the decisive role of consumption in bringing about change in the spatial dynamics of the agri-food system thanks to the role of retailers in responding to new consumer requirements.

But in the "global networks of places", production and processing places can also affect change on places of consumption. This is possible in that social change encourages innovation on the part of a new class of entrepreneurs. These *homines novi* (new men) put forward new ideas for products of two types: (a) new ways of satisfying existing requirements and (b) new products which give rise to new desires in the minds of consumers.

The role of activating the production process passes from one point to another with a certain frequency, so it seems safe to accept the hypothesis that the hierarchy between points on the network is dynamic. This is what makes a given agri-food system competitive.

A further opportunity for the places of production or processing to play a driving role in the spatial dynamics of agri-food systems is given by the increasing concern of consumers about the origin and processing of food products. This demand for food safety is frequently accompanied by awareness of food biodiversity, associated with a non-standardised way of satisfying food requirements.

Spatial dynamics such as those discussed above typify places where agri-food production is based on methods of crop growing, animal farming and processing where human skills are more important than technology (Arfini et al. 2010).

Places (local communities) whose products meet both the need for food safety and biodiversity have a competitive advantage in marketing their products.

In this view, globalisation brings advantages. Local communities can link up more easily thanks to new communications and transport networks.

The fall in the cost of communications has a dual effect. It facilitates on one hand the spread of knowledge of agri-food products among manufacturers and consumers; and on the other the interchange of innovative entrepreneurial ideas between local communities of distant places.

The fall in transport costs also has a double effect. It makes the circulation of agri-food products easier and it encourages the trend towards specialization both in producing and in processing.

Supermarkets can also help to promote the competitiveness of local producers by including selected products in their range.

Consistently with this theoretical framework, research in agri-food appears to confirm that there is greater vitality among producers in places where there is a sense of belonging to the local community. Research shows that globalisation has not led to the disappearance of local distinctions. In fact, given that it is places offering tangible and intangible, cultural and institutional resources, which sustain the innovation and interaction necessary for competition in the global arena, globalisation has often implied a re-evaluation of particular local characteristics (Valdani and Ancarani 2001).

5. Some final remarks

This chapter has suggested that the ID theory can shed new light on the spatial dynamics of agri-food systems, and can offer an alternative to the mainstream approach.

In using the local community as a unit of analysis, ID theory gives a key role to human agents of production and their knowledge. General and specialized human ability, in both technical fields and business, lies at the core of all economic and social change, and therefore the spatial dynamics of production systems, including agri-food systems.

This theoretical position leads us to consider a local community as a learning environment and an incubator of entrepreneurial talent. These two aspects affect the competitiveness of each individual place and of the nation as a whole.

The policy implications of this reasoning are mainly relevant for institutions.

Institutions need to ensure that human resources, the real asset of a place, develop and improve continuously. This should be not only in response to external stimuli from the market; there needs to be a policy of updating knowledge and skills and increasing entrepreneurial capital through support for new enterprises.

Notes

1. See, for instance, *Alternative Agro-Food Networks* (Goodman 2003), *Local Food Systems* (Feenstra 1997), *Short Food Supply Chains* (Renting et al. 2003).
2. In the principal-agent model, all market power lies in the hands of one party, the party offering the contract, who is leader in the vertical hierarchy between producer and distributor. For vertical chain control and the principal-agent model see: Ross (1973); Rey and Tirole (1986a); Rey and Tirole (1986b); Tirole (1989); Salanié (1994); Salanié (1998).
3. Credence attributes are those that the consumer cannot evaluate before sale or during consumption (Nelson 1970).

References

Arfini, F. and Mancini, M.C. (2004). British Retail Consortium (BRC) Standard: a New Challenge for Firms Involved in the Food Chain. Analysis of Economic and Managerial Aspects. In Schiefer, G. and Rickert, U. (eds.), *Quality assurance, risk management and environmental control in agriculture and food supply networks.* Proceedings of the 82nd European Seminar of the European Association of Agricultural Economists (EAAE), 23-31. May 14-16 2003, Bonn, Germany.

Arfini, F., Belletti, G. and Marescotti, A. (2010). *Prodotti tipici e denominazioni geografiche. Strumenti di tutela e valorizzazione.* Gruppo 2013, Quaderni. Roma: Edizioni Tellus.

Barjolle, D., Boisseaux, S. and Dufour, M. (1998). *Le lien au terroir. Bilan des travaux de recherche.* Lausanne: ETH Institut d'économie rurale.

Becattini, G. (1962). *Il concetto d'industria e la teoria del valore.* Torino: Boringhieri.

—. (1979). Dal "settore" industriale al "distretto" industriale. Alcune considerazioni sull'unità d'indagine dell'economia industriale. *Rivista di economia e politica industriale* 5 (1): 7-21.

—. (2004). *Industrial Districts. A New Approach to Industrial Change.* Cheltenham, UK: Edward Elgar.

Belletti, G. and Marescotti, A. (1996). Le nuove tendenze dei consumi alimentari. In Berni, P. e Begalli, D. (eds.), *I prodotti agro-alimentari di qualità: organizzazione del sistema delle imprese*, 133-152. Bologna: il Mulino.

Benoun, M. and Héliès-Hassid, M.L. (2003). *Distribution. Acteurs et Stratégies.* Paris: Economica.

Bérard, L. and Marchenay, P. (1995). Lieux, temps, et preuves: la construction sociale des produits de terroir. *Terrain* 24: 153-164.

Berard, L. and Marchenay, P. (1997). *Rosette et Rosette de Lyon. Eléments historiques et ethnologiques.* Bourg en Bresse: CNRS, Recherche et Information sur les ressources des terroirs.

Bertazzoli, A. et al. (2006). Il marketing nel sistema agro-alimentare tra impresa e struttura di relazioni. In Antonelli, G. (ed.), *Marketing agro-alimentare: specificità e temi di analisi*, 87-116. Milano: Franco Angeli.

Busch, L. and Bain, C. (2004). New! Improved? The Transformation of the Global Agrifood System. *Rural Sociology* 69 (3): 321-346.

Campbell, H., Lawrence, G. and Smith, K. (2006). Audit cultures and the antipodes: the implications of EurepGAP for New Zealand and Australian agri-food industries. In Marsden, T. and Murdoch, J. (eds.), *Between the local and the global: confronting complexity in the contemporary agri food sector*, 69-94. Amsterdam: Elsevier.

Casabianca, F. and Valceschini, E. (1996). *La qualité dans l'agro-alimentaire: émergence d'un champ de recherches. Rapport final AIP Construction social de la qualité.* Paris: INRA-SAD.

Casabianca, F. et al. (2005). *Terroir et typicité: deux concepts-clés des Appellations d'Origine Contrôlée. Essai de définitions scientifiques et opérationnelles.* Communication pour le Symposium international Territoires et enjeux du développement regional, 9-11 Mars 2005, Lyon.

CIRAD (1996). *Systèmes agroalimentaires localisés: organisations, innovations et développement local.* Montpellier: CIRAD-SAR.

Courlet, C. (2008). *L'Économie territoriale.* Grenoble: Presses Universitaires de Grenoble.

Courlet, C. and Pecqueur, B. (1992). Les systèmes industriels localisés en France: un nouveau modèle de développement. In Benko, G. et Liepitz, A. (éds.), *Les régions qui gagnent. Districts et réseaux: les nouveaux paradigmes de la géographie économique*, 81-102. Paris: PUF.

Courlet, C. (2006). Les systèmes productifs localisés (SPL): de que parle-t-on? In Courlet, C. (ed.), *Territoire et développement économique au Maroc: les cas des systèmes productifs localisés*, 15-23. Paris: L'Harmattan.

Daumas, J.-C. (2007). Dans la "bôite noire" des districts industriels. In Daumas, J.-C., Lamard, P. et Tissot, L. (éds.), *Les territoires de l'industrie en Europe (1750-2000). Entreprises, régulations et trajectoires*, 10-34. Besançon: Presses Universitaires de Franche-Comté.

Davis, H.J. and Goldberg, R.A. (1957). *A Concept of Agribusiness.* Cambridge: Harvard University Press.

De Sainte Marie, C. (1996). *Spécification des produits et appropriation juridique. La qualité dans l'agroalimentaire, émergence d'un champ de recherches.* Paris: INRA.

—. et al. (1995). La construction sociale de la qualité. Enjeux autour de l'Appellation d'Origine Contrôlée "Brocciu Corse". In Nicolas, F. et Valceschini, E. (éds.), *Agro-alimentaire: une économie de la qualité.* Paris: INRA Editions-Economica.

Farina, E. et al. (2005). Determinants of retailers' decisions to use public or private grades and standards: Evidence from the fresh produce market of São Paulo, Brazil. *Food Policy* 30 (3): 334-353.

Feenstra, G. (1997). Local food systems and sustainable communities. *American Journal of Alternative Agriculture* 12 (1): 28-36.

Fulponi, L. (2006). Private voluntary standards in the food system: the perspective of major food retailers in OEDC countries. *Food Policy* 31 (1): 1-13.

—. et al. (2006). Sécurité sanitaire et normes collectives de distributeurs: impact sur les filères et l'offre alimentaire. *INRA Sciences Sociales* 5-6: 1-4.

Galizzi, G. (1975). L'emergente sistema agricolo alimentare. *Agricoltura* 24 (1): 36-52.

Ghersi, G. and Bencharif, H. (1995). Sistemi e offerta agro-alimentari. In Malassis, L. e Ghersi, G. (eds.), *Introduzione all'economia agro-alimentare*, 97-189. Bologna: il Mulino.

Giacomini, C., Mancini, M.C. and Modesti, P. (2010). Large retailers and vertical control of the supply chain: private labels and private certification standards. *International Journal of Applied Management Science* 2 (4): 305-320.

Giraud-Hèraud, E., Hammoudi, H. and Soler, L.G. (2006). *Food Safety, Liability and Collective Norms.* École Polytecnique, Laboratoire d'Économétrie. Cahier 6.

Goodman, D. (2003). The quality "turn" and alternative food practices: reflections and agenda. *Journal of rural studies* 19 (1): 663-687.

Grossman, G.M. and Rossi-Hansberg, E. (2006). The rise of offshoring: It's not wine for cloth anymore. Paper prepared for the Symposium on The New Economic Geography: Effects and Policy Implications, Federal Reserve Bank of Kansas City, Jacksonhole, Wyoming, 24-26 August.

Grossman, G.M. and Rossi-Hansberg, E. (2008). Trading Tasks: A Simple Theory of Offshoring. *American Economic Review* 98 (5): 1978-1997.

Hammoudi, A., Hoffman, R. and Surry, Y. (2009). Food safety standards and agri-food supply chains: an introductory overview. *European Review of Agricultural Economics* 36 (4): 469-478.

Hatanaka, M., Bain, C. and Busch, L. (1996). Differentiated standardization, standardized differentiation: the complexity of the global agri-food system. In Marsden, T. and Murdoch, J. (eds.), *Between the local and the global: confronting complexity in the contemporary agri-food sector*, 39-68. Amsterdam: Elsevier.

Henson, S. and Reardon, T. (2005). Private agri-food standards: implication for food policy and the agri-food system. *Food Policy* 30 (3): 241-253.

Hirst, P. and Thompson, G. (2003). The future of globalisation. In Michie, J. (ed.), *The Handbook of Globalisation*, 17-36. Cheltenham, UK: Edward Elgar.

Letablier, M.T. and Delfosse, C. (1995). Genèse d'une convention de qualité. Cas des appellations d'origine fromagères. In Allaire, G. et

Boyer, R. (eds.), *La grand transformation de l'agriculture*, 97-118. Paris: INRA Editions-Economica.

Loader, R. and Hobbs, J.E. (1999). Strategic responses to food safety legislation. *Food Policy* 24 (6): 685-706.

Malassis, L. and Bourdon M. (1970). Un modèle simple de développement agricole intégré. *Cahiers de l'ISEA* IV (2).

MITYC (2006). Orden ITC/2691/2006, de 2 de agosto, por la que se regulan las bases, el régimen de ayudas y la gestión de medidas de apoyo a las agrupaciones empresariales innovadoras. BOE 199 Lunes 21 agosto 2006: 30854-30862.

Muchnik, J. (2010). Localised agri-food systems: concept and diversity of situations. *Sviluppo Locale* 14 (35): 3-20.

Nelson, P. (1970). Information and consumer behaviour. *Journal of political economy* 78 (2): 311-329.

Porter, M.E. (1990). *The Competitive Advantage of Nations*. New York: The Free Press.

Porter, M.E. and Ketels, C. (2009). Clusters and industrial districts: common roots, different perspectives. In Becattini, G., Bellandi, M. and L. De Propris (eds.), *The Handbook of Industrial Districts*, 172-183. Cheltenham, UK: Edward Elgar.

Renting, H., Marsden, T.K. and Banks, J. (2003). Understanding alternative food networks: exploring the role of short food supply chains in rural development. *Environment and Planning* A (35): 393-411.

Requier-Desjardins, D. (2007). *L'évolution du débat sur les SYAL: le regard d'un économiste*. XLIIIè colloque de l'ASRDLF, Les dynamiques territoriales: débats et enjeux entre les différentes approches disciplinaires. Grenoble et Chambéry, 11-13 Juillet.

Rey, P. and Tirole, J. (1986a). Contraintes verticales: l'approche principal-agent. *Annales d'Économie et de Statistique* 1: 175-201.

Rey, P. and Tirole, J. (1986b). The logic of vertical restraint. *The American Economic Review* 76 (5): 923-939.

Ross, S.A. (1973). The Economic Theory of Agency: The Principal's Problem. *American Economic Review* 63 (2): 134-139.

Salanié, B. (1994). *Théorie des contracts*. Paris: Economica.

—. (1998). *Microéconomie. Les défaillances du marché*. Paris: Economica.

Sylvander, B. (1995). Conventions de qualité, concurrence et coopération. Cas du "label rouge" dans la filière Volailles. In Allaire, G. et Boyer, R. (éds.), *La grande transformation de l'agriculture*, 167-183. Paris: INRA Editions/Economica.

—. (1996). Les controverses sur la politique de la qualité: les dispositifs d'ajustement et la certification. In Casabianca, F. and Valceschini, E. (eds.), *La qualité dans l'agro-alimentaire: émergence d'un champ de recherche*, 106-119. Paris: INRA-SAD.

Sylvander, B. and Lassaut, B. (1994). L'enjeu économique de la qualité sur les marchés des produits agro-alimentaires. In Multon, J.L. (éd.), *La qualité des produits alimentaires: politique, incitations, gestion et contrôle*. Paris: Lavoisier.

Tirole, J. (1989). *The Theory of Industrial Organization*. Cambridge: The MIT Press.

UNIDO (2001). The UNIDO Cluster/Network Development Programme: The Italian Experience of Industrial Districts. Published online http://www.unido.org/index.php?id=o4310, accessed 24 June 2011.

Valceschini, E., Saulais, L. et Barrey, S. (2005). *Articulation entre réglementation, normalisation et référentiels privés dans les industries agro-alimentaires. Étude sur le système de standardisation de la sécurité alimentaire dans le IAA*. Paris: DPEI-ADEPRINA, Ministère de l'Agriculture.

Valdani, E. and Ancarani, F. (2000). *Strategie di marketing del territorio. Generare valore per le imprese e i territori nell'economia della conoscenza*. Milano: Egea.

CHAPTER TWO

SUSTAINABLE INTER-ORGANIZATIONAL RELATIONSHIPS IN REGIONAL AND NON-REGIONAL AGRI-FOOD SUPPLY CHAINS

CHRISTIAN FISCHER

Abstract

"Local" and "regional food" has received increased attention from different social groups as it is thought to provide environmental and economic benefits over "global" or "non-regional food". While briefly discussing these issues the key question behind this study is how supply chain managers deal differently with more distant (i.e., non-regional) suppliers or buyers than with closer (i.e., regional) ones? Using a large company survey dataset, different structural equation models are estimated. The results support the assumption of potentially existing structural differences in supply chain management requirements between the two types of agri-food chains. Regional chain relationships are shown to last longer and they are more dependent on personal bonds. Such relationships are also slightly more susceptive to an equal power balance between partners. Hence, regional chain relationships, similar to the ones with farmers, need to be managed in principle based on personal contacts rather than via effective but more or less remote communication. The main implication arising from this study is that sustainably increasing the supply of locally produced foods on a large-scale basis will require the enterprises involved (farmers, processors, retailers) and support organisations to interact and transact in a certain, highly effective way.

1. Introduction

"Local" food has recently received increased attention from consumers, policy-makers and researchers (Martinez et al. 2010). While the term "local" is imprecise, the demand for such food products has nevertheless grown. "Local is the new organic" (Cloud 2007), a concept as fashionable but maybe less ideologically or politically loaded. Most people can identify themselves with local products, be it as a citizen (from "my home or birthplace") or as a tourist (enjoying "regional specialities"). Can there be anything wrong with buying and consuming products that require little shipping and storage, and come from producers that one knows and can visit and thus trust?

In a world that is organised according to production efficiency, yes. International specialisation in production (country A produces wine, country B clothing etc.) and cross-border trade can exploit economies of scale. And as long as such production cost gains exceed logistics expenses (transportation and warehousing) it makes economic sense to engage in the division of labour and the international exchange of goods and services. Moreover, producing at the global cost minimum makes everyone richer (or at least the world as a whole) and everyone (or at least the winners from globalisation) can consume more. So why produce potatoes in every country if they can all come from a few nations specialising in growing them?

The problem with internationally specialised agricultural production is that monocultures deplete soils, threaten biodiversity, concentrate risks (it is an anti-diversification strategy) and that conventional large-scale, long-distance shipping consumes significant amounts of energy, space (for logistics infrastructure) and causes harmful combustion emissions and noise. Properly accounting for all these "externalities" and factoring them into the total cost bill, international production specialisation becomes much less attractive. So shipping potatoes from country A to country B no longer makes economic sense if overall specialisation costs exceed the gains from production economies of scale. Better then growing potatoes wherever possible, being able to visit producers and look them over the shoulder to ensure they do a proper job,, feel good about the support to the local community and pay a "fair" price, i.e., one that does not "exploit" anyone and helps to protect the environment. These are the reasons why local food is "in".

Unfortunately, the world is not black and white, but rather a spectrum of shades of grey. Some globally appreciated food products (e.g., coffee, cacao, tropical fruits, olive oil) cannot reasonably be produced everywhere and thus must be traded. Moreover, food production in general is seasonal, and for some perishables, such as fruit and vegetables, extra-seasonal international trade can make sense in cases where storage costs (for local produce) exceed transportation costs (for imported produce). Furthermore, since the job of food producer is not universally considered as highly attractive, there may simply not be enough local food available in all places so that food needs to be procured (imported) from areas where it is more plentiful. Finally, conscious consumption of "exotic" foods may help economic progress in developing countries thus making it sometimes meaningful for industrialised-country citizens to favour, for instance, organic unrefined artisan cane sugar over local industrial white beet sugar. So it is likely that the world's food will be continued to be produced wherever feasible and traded whenever necessary. In the future, the only difference may be that more thinking may go into when to do what, and such justifications will need to be made explicit, resulting in more campaigning, branding and labelling.

Given the fact that local and international food supply chains will continue to coexist—and both types will be important in the future—one question that arises is whether there are structural differences in the management practice of these chains. That is, do managers in one type of the food supply chain need to adopt a different approach than in the other? Such differences may exist in many areas, finance, logistics, human resources, marketing etc. In this chapter, the focus is on the management of the personal interactions between supply chain buyers and sellers, i.e., their collaboration practices.

The analysis which follows is based on a large dataset that was collected during an EU Framework 6 research project. The project investigated the nature and determinants of sustainable inter-organizational relationships (SIRs) in selected European agri-food chains. The results of this work have been published widely and are briefly reviewed in this chapter. New empirical insights concerning the differences in relationship sustainability in regional and non-regional chains are added and the results are then discussed in the light of the earlier work. In addition to the differences in agri-food chain relationships differentiated by geographical distance, empirical estimates differentiated by relationship age (or "maturity") are also presented. These results complement an earlier study (Reynolds et al. 2009) that was conducted on a subset of the entire database. Overall, the results presented in this chapter provide further empirical evidence on some specific aspects of agri-food supply chain management on which no extensive recent literature exists.

This chapter is structured as follows. After a brief summary of the relevant literature, research hypotheses are formulated and the empirical approach (data and methodology) is explained. The results are then presented and discussed and, finally, some conclusions are drawn.

2. Local, regional and beyond: does distance matter for collaboration practice in agri-food supply chains?

The concepts of "local" and "regional" are similar but not identical and mean different things to different people. Nevertheless, "local" is sometimes seen as an antipode to "global" (Hinrichs 2003) and thus often used by people and movements opposed to, or sceptical of, globalisation and/or its consequences. "Regional", on the other hand, is less politically sensitive and perhaps a "clearer conceptual terminology" (Donald et al. 2010, 174). However, Kneafsey (2010) argues that the term "region" is also ambiguous. For instance, Europe is sometimes referred to as a (macro-)region and yet consists itself of many nations and (sub-)regions.

Generally speaking, a "region" is a distinctive geographical space, or territory, which can be differentiated in some way (culture, language, history, topology, vegetation, climate etc.) from its surrounding areas. A region may also be seen as a human (or social) construct that gives identity to the people living in it. Defined in this way, a region is not the same as a government-specified administrative territorial unit (e.g., district, county, department, province, state) although such territories may well coincide with a "region" as a distinctive geographic space some people identify themselves with. As with most social constructs, a region (its boundaries) must be defined by people and, given that individuals do not usually all agree, the definition of a region is then also the result of a majority agreement. Thus, the exact definition of a region can differ from person to person and may evolve over time. Using this approach to "region" as a social construct it is not difficult to see what the antipode of the concept is. If a "region" is a synonym for "we" then the opposite is probably "the others" (i.e., those living outside the "region"). Hence, there is only "regional" and "non-regional" (i.e., not in "our region"). Only if the term "region" refers to a government-specified administrative territory, an ordering of terms such as "regional", "national" and "international" is meaningful.

The definition of a "local" or "regional food product" is often more straightforward. The literature describes a local or regional food product as one that is consumed within some area surrounding its production location. Distances are arbitrarily defined but 75, 100, 200, 400 or 500 miles or kilometres are commonly used (Martinez et al. 2010). On the other hand, a "locality food" is one that is produced in a certain area and marketed with an origin designation but is largely consumed elsewhere, e.g., citrus fruit from Sicily or Florida (ibid.). Whatever the exact definition of "local" or "regional" foods, the core difference between them and "non-local" or "non-regional" ones is in the geographic distance between the locations of production and consumption.

In other words, the issue is a "short" food supply chain versus a "long" one. "Short" and "long" supply chains can also be defined in terms of actors or stages involved in them. In this context, "short" means, for example, the direct sales of food from a farmer to the consumer. Of course, such direct sales can also occur across large geographic distances (e.g., online selling). But in this chapter "short" and "long" are used to differentiate spatial proximity.

Distance ("proximity", "closeness" or "propinquity" are related terms used in the literature) has several dimensions. Sometimes there is a distinction between geographic and social distance (e.g., Sorenson 2003). A more differentiated approach also considers cultural, administrative and economic components of distance (Ghemawat 2001). Geographic distance refers to the fact that there is a significant physical obstacle to surmount when exchanging goods or services between locations. Such obstacles cannot always be measured in miles or kilometres. In some cases, natural barriers such as water (sea, lakes, rivers), mountains, forests etc. may separate locations from each other, thus making the exchange of goods or services between them costly in terms of money, time and/or adverse environmental impacts. But in general, geographic distance refers to long transport distances. Other dimensions of distance more or less describe differences in business environments such as production methods, law and regulations, general economic conditions, culture and language, climate etc. Ghemawat (2001) shows that for processed food products, cultural distance is the main impediment to international trade while geographic distance is more relevant for perishable products such as fresh fruit.

When business conditions differ it can generally be expected that trading partners have less control over commercial transactions compared to the ones in short supply chains, which potentially results in higher business risks (Fischer 2004). As a consequence, it can be assumed that partnerships are managed differently in long supply chains compared to short ones and that business relationships in these partnerships may also differ in some way. In other words, the key question behind the study described in this chapter is how supply chain managers deal differently with more distant suppliers or buyers than with closer ones.

3. Research hypotheses

The role of geographic location in the human production and distribution of goods and services has been investigated by the economics, management and marketing disciplines (for surveys of the respective literatures see, for instance, Sorenson 2003; Ghemawat 2001 and Ganesan et al. 2005). This chapter reviews only those aspects of economies (and other advantages) of co-location that are relevant for inter-organizational and inter-personal commercial transactions and interactions within vertical business partnerships. The literatures discuss three areas in which distance affects inter-organizational and/or inter-personal business trans- and interactions: (i) the durability of the relationships; (ii) the role of personal ties (or bonds) within such relationships; and (iii) the nature of communication between supply chain partners.

3.1 The durability of relationships

According to Sorensen (2003), people can be expected to interact more frequently with those who live in close geographic proximity and to whom they are culturally close. This is because proximity increases the likelihood of a chance encounter and influences the probability of forming and maintaining a relationship. In order to start and build a relationship, people typically must meet in space and time. Moreover, it can generally be assumed that the durability or maturity of relationships (i.e., their "age") is increased because proximity reduces the costs and other obstacles of maintaining a relationship (Cannon and Homburg 2001). Hence, in short supply chains business relationships should last longer.

> *Hypothesis 1a*: The average length (or maturity) of buyer–supplier relationships in regional (i.e., comparatively short) agri-food supply chains can be expected to be longer.

Long-lasting agri-food chain relationships can be seen as a two-dimensional construct consisting of a static and a dynamic sustainability component (Fischer and Reynolds 2010). The static component represents quality aspects such as the current levels of trust, satisfaction and commitment existing between two business partners. The dynamic component reflects the fluctuations of these "qualities" over time and is referred to as relationship stability. A relationship can only be "sustainable" if it is characterised by sufficiently high levels of quality and stability. If nothing changes, the relationship can be expected to last a long time. However, while it is true that a long-lasting relationship must have been and currently is sustainable, it cannot be assumed with certainty that a currently sustainable relationship is necessarily long-lasting. Nevertheless, in general, one could expect more mature (i.e., older) relationships to be more sustainable.

> *Hypothesis 1b*: The quality and stability ("sustainability") of agri-food chain relationships should increase with maturity.

While spatial proximity provides better opportunities to form close and sustainable relationships this does not mean that all relationships between companies which are located at a short distance to each other are automatically close and sustainable. For instance, Ganesan et al. (2005) show in their case study of successful new-product development that economies of co-location can only be expected if "strong relational ties" exist, indicating that good, sustainable relationships are necessary for industrial cluster success but cannot be taken for granted in spatial agglomerates. Nevertheless, if relationships are expected to last longer in regional business partnerships (Hypothesis 1a) and longer-lasting ones are assumed to be more sustainable (Hypothesis 1b), then regional chain relationships, on average, should also be more sustainable.

> *Hypothesis 1c*: The quality and stability ("sustainability") of agri-food chain relationships should, on average, decrease with distance.

3.2 The role of personal ties or bonds

Business relationships are typically formal relationships governed by cost–benefit considerations, but personal ties or bonds (i.e., friendship) can also develop between business partners. The significant role of such personal bonds in agri-food chain relationships has been empirically demonstrated, in particular for farmers (Fischer et al. 2010). Sorenson (2003, 515) argues that geographic proximity facilitates the "maintenance of social relationships […], particularly close personal ties". Ganesan et al. (2005, 44) also state that it is commonly assumed that "geographic proximity […] serves to enhance […] the development of strong personal ties". Hence, spatial, cultural and social closeness contributes to fostering personal bonds, in private as well as business life.

> *Hypothesis 2*: Personal bonds or ties should play a more important role in regional business relationships as compared to non-regional ones.

3.3 The nature of communication

Large geographic distances can hamper communication between business partners. Cannon and Homburg (2001) cite time zone differences as a potential barrier. Language problems may be another issue (Ghemawat 2001). Moreover, geographic proximity is expected to enhance face-to-face communication (Ganesan et al. 2005), based on the fact that personal encounters (e.g., business meetings) are easier and less costly to arrange (Sorenson 2005). However, as Ganesan et al. (2005) show, in business, face-to-face communication is not necessarily superior to other types of communication (e.g. telephone or e-mail) and interpersonal communication is only more frequent and crucial in such business partnerships which are built on strong relationships. On the other hand, it is clear that successful coordination of commercial transactions and interactions across long distances is more complex and that it requires the effective exchange of information, impersonal (i.e., by traditional as well as electronic means) and personal (i.e., communication). In fact, given that in long-distance business partnerships personal meetings and visits are more difficult and costly and thus probably less frequent, effective communication between these partners becomes more crucial. Good communication serves to build trust and trusting business partners are more likely to be more committed and satisfied (Garbarino and Johnson 1999). Hence, as geographic distance increases, so does the importance of effective communication.

Hypothesis 3: Effective communication is more important in long-distance business partnerships.

4. Empirical approach for testing the research hypotheses

Differences in buyer–supplier relationships between regional and non-regional agri-food chains are examined using a dataset that was collected in six different EU countries (Germany, UK, Spain, Ireland, Finland and Poland), for two different commodities (meat and cereals) and two chain stages (upstream: farmer–processor and downstream: processor–retailer). Some information on the data and analysis methodology is provided below, but for more details on both see Fischer et al. (2009) and Fischer et al. (2010).

4.1 Survey data

Based on pilot study findings involving expert interviews (Fischer et al. 2007), a company survey was developed. The questionnaire was pre-tested separately in each participating country. Where feasible, personal interviews were conducted (mostly with farmers) or respondents were interviewed by telephone. In addition, questionnaires were sent by mail and followed up by telephone calls and/or a subsequent reminder mailings.

Most of the obtained samples were drawn from existing sampling frames, i.e., all businesses whose contact details were available were approached.[1] The data were collected from November 2006 to April 2007. In total, the surveys yielded 1,442 usable responses (see Fischer et al. 2010 for sample profiles).

The representativeness of the convenience samples was assessed using two criteria for which complete target population information was available across the different countries: geographic distribution of farm/company locations and farm/company size. Despite some differences across the countries,[2] overall the collected responses reflect existing population variations with regard to these two indicators.

As to other aspects of data quality, in total 86% of survey respondents claimed to be in upper management positions or (part-) owners of the businesses surveyed. Non-response bias was assessed by comparing early survey responses with later ones, using multivariate analysis of variance on key demographic characteristics (company/farm size, geographic location and business activity). Again, no significant differences were found.

4.2 Statistical methods used

The relevance of different factors potentially affecting the sustainability of agri-food chain relationships were examined using structural equation modelling (SEM).[3] SEM in its most general form consists of a set of linear equations that simultaneously test associations between directly observable and/or latent variables that cannot directly be measured (Bollen 1989).

All SEMs discussed in this chapter were estimated as multi-group models, keeping the model specification constant for all groups.

The underlying generic model performed best on theoretical and statistical grounds for the overall (pooled) dataset. One of the drawbacks of this approach is that a generic model may not necessarily be the best specification for each sub-group (country, commodity chain, chain stage, etc.). Nevertheless, a generic SEM is most appropriate because the aim of the analysis in this chapter is to compare the relevance of fundamental determinants of agri-food relationship sustainability between regional and non-regional agri-food chains, and young and mature relationships.

The generic model specification is shown in Fig. 2-1. Based on theoretical work (see Fischer et al. 2009 and Fischer and Hartmann 2010), and a qualitative study (Fischer et al. 2007), the construct of relationship sustainability is assumed to be a function of partnership-level factors such as effective communication, the existence of personal bonds, the impact of key people leaving the partnership, and equal power distribution among chain partners. For descriptive statistics of these variables see Fischer et al. (2009).

Two constructs (multi-item, latent variables) were used in the specified model. All other variables were measured as single items. In the following, when referring to the variables used in the SEM, their names will be capitalised. The Relationship Sustainability construct was specified as a one-level, four-item latent variable.[4]

For the generic model, construct reliability was assessed by Cronbach's alpha. Construct validity was assessed using principal component analysis (PCA) on the four items. For the Effective Communication construct, only two items were used: Adequate Communication Frequency and High Information Quality. As there are only two items, PCA is not meaningful for this construct.

Finally, despite great care undertaken in questionnaire design and item wording to minimise the problem, common method bias was assessed (i.e., whether significant measurement error due to multiple-item questions limited the validity of the results). No such bias was found. Taken together, these statistical tests suggest that there are no major problems with either the reliability or validity of the two constructs, or the model specification.

Fig. 2-1 Determinants of Relationship Sustainability—the generic model

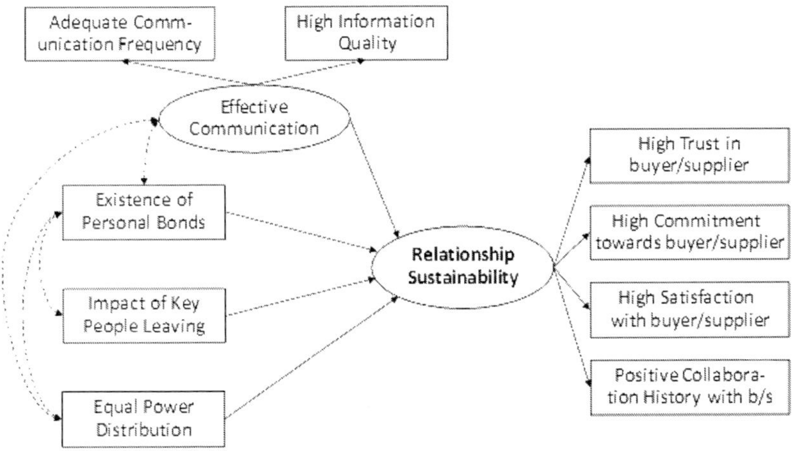

5. Results

Descriptive statistics and mean-difference tests were used in addition to SEM results to provide empirical evidence for the research issues under investigation in this chapter. The classification of their main business relationship as regional or non-regional was left to the respondent without providing distance thresholds. Given that the concept of a "region" is relative and subjective (see discussion above), it is theoretically and practically difficult to pre-define a region in a company survey that covered different EU countries. As to the measurement of the maturity of their most important buyer or supplier relationship, respondents were asked to indicate its "age" in years and months. The result tables below show only those survey results where data was provided on either of these two measures, which explains the difference in observation numbers.

Table 2-1 provides information on the differences in relationship maturity between regional and non-regional agri-food chains. On average, regional chain relationships have lasted for almost 13 years while those in non-regional chains have an age of almost 8.5 years. The difference is statistically highly significant. This confirms Hypothesis 1a which states that relationships can be expected to last longer between business partners that are more closely located to each other.

Table 2-1 Mean maturity (in years) of business relationships in selected EU agri-food chains differentiated by type of agri-food chain

Type of agri-food chain	Mean relationship maturity (in years)	Standard deviation	Number of observations (n)	Statistical significance of difference (F test)
Regional	12.8	11.9	515	F = 27.385
Non-regional	8.4	7.0	233	p = 0.000

Table 2-2 reports Relationship Sustainability index scores for the different sub-groups which were obtained when differentiating the overall dataset according to geographic distance and relationship maturity. The index scores are unweighted mean values which were calculated on the basis of four individual components (items) shown in Fig. 2-1 ("Our trust in this supplier/buyer", "Our commitment towards this buyer/supplier", "Our satisfaction with this buyer/ supplier" and "Our collaboration with this buyer/supplier in the past"). Each item was measured on a rating scale reaching from 1 = "very poor" to 7 = "very good". On the other hand, relationship maturity was measured continuously, and in order to perform a multi-group analysis, discrete sub-groups needed to be created. The mean age of the relationships contained in the dataset is 11.3 years and the median one 8.7 years. Although this was a more or less arbitrary decision, a cut-off age of 10 years was used to distinguish young relationships from mature ones.

As the results in Table 2-2 indicate, Relationship Sustainability index scores are significantly higher in more mature relationships than in younger ones. In absolute terms, the difference between the two groups is only small. Moreover, the association between the two variables proved to be weak when measured in different ways. For instance, correlation coefficients between the two variables are positive and statistically significant at least at the 95% confidence level, but low (Pearson: 0.09; Spearman: 0.13). Nonetheless, although the data only provide weak support, Hypothesis 1b, which states that more mature relationships should be more sustainable, can be confirmed. On the other hand, Hypothesis 1c, stating that, on average, business relationships in regional as compared to non-regional supply chain partnerships should also be more sustainable needs to be rejected. As the results show, even if in absolute terms Relationship Sustainability index scores are somewhat higher in regional relationships, the difference between the two types of agri-food chains is not statistically significant.

Table 2-2 Sustainability levels* of business relationships in selected EU agri-food chains, differentiated by geographic distance and relationship maturity

Analysed sub-group	Relationship Sustainability* index score (mean)	Standard deviation	Number of observa-tions (*n*)	Statistical significance of difference (*F* test)
Relationship maturity				
• young (< 10 years)	5.64	0.93	378	$F = 6.228$ $p = 0.013$
• mature (≥ 10 years)	5.81	0.92	353	
Geographic distance				
• regional	5.63	0.92	901	$F = 0.345$ $p = 0.557$
• non-regional	5.59	0.90	355	

Notes: *Mean values of index score calculated on the basis of four individual components (items), each one measured on a rating scale (1 = very poor, ..., 7 = very good). The items are "Our trust in this supplier/buyer", "Our commitment towards this buyer/supplier", "Our satisfaction with this buyer/supplier" and "Our collaboration with this buyer/supplier in the past". The index scores are unweighted averages of the four obtained item scores. Only cases with valid observations on all four items were considered.

Table 2-3 provides the SEM results. In order to be better able to interpret the new estimates for the maturity and geographic-distance sub-groups, previously obtained SEM estimates are also listed. Only a selection of the results is reported, however. For instance, the bivariate correlation paths between the regressors and the measurement model estimates are not mentioned but can be found for the previously published results in the cited sources.

As to the relationship maturity estimates, for young relationships only two determinants are relevant: Effective Communication and the Existence of Personal Bonds, with the former being significantly more important than the latter.[5] In more mature agri-food chain relationships, the Existence of Personal Bonds is equally important to Effective Communication. In addition, in these relationships the Impact of Key People Leaving a supply chain partnership has some (negative) effect on the underlying relationship, as has Equal Power Distribution (positive impact) between the partners. The overall model fit is slightly less satisfactory than the other previously estimated SEMs. Nevertheless, with a CMIN/DF of just below 2 (and an NFI of 0.94 and a RSMEA of 0.035) the specification can still be seen as highly acceptable. About 50% of the variance of the

Relationship Sustainability construct is explained in these models (a little bit less in the "young" sub-group, a little bit more in the "mature" one), which is in line with the previously estimated models and a good result considering the cross-sectional nature of the data.

Turning to the geographic distance-specific results, regional agri-food chain relationships are characterised by the influence of all specified determinants, with the impact of Effective Communication being highest, followed by the Existence of Personal Bonds between supply chain partners, Equal Power Distribution between them and a small negative Impact of Key People Leaving the partnership. These findings confirm Hypothesis 2 which postulates the special importance of personal bonds in regional agri-food chains. In non-regional relationships just three determinants are important for Relationship Sustainability: Effective Communication, the Existence of Personal Bonds and Equal Power Distribution. Interestingly, here the influence of Effective Communication is extraordinarily high, in fact it is the highest across all previously estimated models, thus confirming Hypotheses 3 which postulates just that. The impact of the Existence of Personal Bonds and of Equal Power Distribution is almost similar in magnitude. As to the model fit, with a CMIN/DF of 1.45 (and an NFI of 0.99 and a RMSEA of 0.018) it is better than the relationship-maturity specification but still not as good as some of the other previously estimated models. The R^2 of the Relationship Sustainability construct is about 50% as in all previously estimated SEMs.

Table 2-3 Summary of the importance of Relationship Sustainability determinants and overall model statistics for the aggregate dataset and different sub-groups including geographic distance and relationship maturity

Model specification	Structural model (standardised path coefficients)				Overall model statistics		
	Effective Comm-unication	Existence of Personal Bonds	Impact of Key People Leaving	Equal Power Distribution	CMIN/DF (p)	R^2 of key construct "relationship sustainability"	n
Aggregate model (Fischer et al. 2009)	0.51***	0.26***	–0.06**	0.16	1.479* (0.069)	0.484	1,442
Contract type-specific estimates (Fischer et al. 2009)							
• implicit	0.52***	0.23***	–0.08**	0.16***	1.239 (0.133)	0.479	852
• explicit	0.56***	0.29***				0.588	289
Country-specific estimates (Fischer et al. 2010)							
• Germany	0.29***	0.38***	–0.11*	0.42***	1.336*** (0.006)	0.625	140
• UK	0.55***	0.15**		0.18**		0.543	254
• Ireland	0.32***	0.36***	–0.18*			0.414	151
• Finland	0.37***	0.30***	–0.11**	0.32***		0.608	226
• Poland	0.66***	0.21***				0.551	335
• Spain	0.48**			0.24***		0.392	336
Commodity-specific estimates (Fischer et al. 2010)							
• meat	0.55***	0.26***	–0.07**	0.12***	1.221 (0.149)	0.512	967
• cereals	0.46***	0.23***		0.20***		0.428	463

–continued–

Model specification	Structural model (standardised path coefficients)				Overall model statistics		
	Effective Comm-unication	Existence of Personal Bonds	Impact of Key People Leaving	Equal Power Distribution	CMIN/DF (*p*)	R^2 of key construct "relationship sustainability"	*n*
Chain stage-specific estimates (Fischer et al. 2010)							
• farmer–processor	0.49***	0.29***	–0.07**	0.15***	1.349* (0.061)	0.475	1,086
• processor–retailer	0.62***	0.10*		0.14**		0.499	344
Relationship maturity-specific estimates (compare Germany-specific results in Reynolds et al. 2009)							
• young (< 10 years)	0.55***	0.24***			1.979*** (0.000)	0.474	414
• mature (≥ 10 years)	0.42***	0.41***	–0.09*	0.16***		0.520	368
Geographic distance-specific estimates							
• regional	0.44***	0.29***	–0.07**	0.19***	1.451** (0.027)	0.456	954
• non-regional	0.68***	0.13**		0.09*		0.586	379

Notes: significance levels: *** (p < 0.01), ** (p < 0.05), * (p < 0.10); non-significant values are not reported.
Source: author's estimates and above mentioned articles.

6. Discussion and conclusions

The hypotheses that regional relations ought to be, on average, longer lasting and more sustainable, in particular when also assuming more mature relationships to be more sustainable, could not be fully confirmed. The results from the available dataset show that regional agri-food relationships, on average, are not more sustainable than non-regional ones. This may simply be a data problem, considering that Relationship Sustainability scores are actually higher in regional supply chain partnerships, but the difference is too small to be statistically (and practically) significant. Further research, involving other data, would be useful to bring more clarity into this issue.

As to the determinants of Relationship Sustainability, the estimated effects are fairly similar in regional agri-foods chains to those in farmer–processor relationships and those in the meat supply chain (see Table 2-3). Such supply chains have in common that business procedures and product standards in general tend to be less developed and formalised than in non-regional, processor–retailer and cereal chain relationships. Trust may be more of an issue in the relationships of the former group, so there could be a need for a stronger personal component in the commercial trans- and interactions. Obviously, this is a general finding from the available data and while it seems to hold on average, the situation in individual cases (i.e., supply chains) may well be different.

The especially crucial role of good, effective communication in non-regional agri-food chains clearly confirms the theoretical expectations. As distance increases and the coordination of business trans- and interactions becomes more challenging, more and better information needs to be exchanged and communication grows in importance. Although not all non-regional business partnerships experience time zone differences and language problems, some international supply chains fall into this category and these obstacles frequently need to be dealt with. Effective communication is important in many areas of life, and it is particularly critical in long-distance relationships.

Regarding relationship maturity, the analysis provides clear evidence that regional agri-food relationships are on average significantly older. Moreover, it seems that the more mature a supply chain relationship, the more important personal bonds become, and the relative significance of effective communication decreases. It is obvious that as time passes, there are more opportunities to get to know each other and that friendships can develop. In fact, it is probably unlikely that business relationships in which partners do not get along well with each other grow very old. Equally, as each partner gets to know, learns about, and increasingly relies on the other, communication intensity and needs probably decline. These results are in line with the findings of Reynolds et al. (2009), thus confirming these using a considerably larger dataset.

To summarise, the results support the assumption of potentially existing structural differences in supply chain management requirements between regional (i.e., "short) and non-regional (i.e., "long") agri-food chains. Regional chain relationships have been shown to last longer and they are more dependent on personal bonds. Consequently, if important relationship stakeholders leave, the impact of this disruption is felt. Such relationships are also slightly more susceptive to an equal power balance between partners. Hence, regional chain relationships, similar to the ones with farmers, need to be managed in principle on the basis of personal contacts rather than via effective but more or less remote communication.

The main implication arising from this study is that local or regional food should not only receive attention from consumers, policy-makers and researchers (Martinez et al. 2010) but that the increased demand for it also has consequences for management practice. Sustainably increasing the supply of locally produced foods on a large-scale basis will require the enterprises involved (farmers, processors, retailers) and support organisations to interact and transact in a certain, highly effective way. The "best practice" approach in this regard still has to be identified, developed and disseminated by research and educational organisations and also to be adopted and applied by practitioners. The training of the next generation of food and agribusiness managers will need to be extended accordingly.

Future research may be directed at gaining a better understanding of other management requirements in local food supply chains. This study has made a first contribution and demonstrated that vertical relationship management practices should give attention to certain partnership aspects. Needless to say, there is ample scope for further work. Potentially valuable areas for further investigations with practical food supply chain management implications could be whether direct sales of local or regional food from producers to consumers are preferable to marketing via distribution and retailing "middlemen", whether local food production in urban and peri-urban settings should be centralised or rather locationally spread to be even closer to local consumers, and/or to what extent willing city and town dwellers could contribute to local food supply by growing vegetables and rearing animals on balconies, roof tops and in private gardens.

Notes

The data used for the empirical analysis of this chapter were collected in the research project on "Key factors influencing economic relationships and communication in European food chains" (FOODCOMM, SSPE-CT-2005-006458) which was funded by the European Commission as part of the Sixth Framework Programme. The collaborating laboratories were: Institute for Food and Research Economics, University of Bonn, Germany (coordinator); Land Economy and Environment Research Group, Scottish Agricultural College (SAC), Edinburgh and Aberdeen, UK; Institute for Agricultural Development in Central and Eastern Europe (IAMO), Germany; The Ashtown Food Research Centre (AFRC), Teagasc, Dublin, Ireland; Ruralia Institute, University of Helsinki, Finland; Institute of Agricultural and Food Economics (IAFE), Poland; and Unit of Agri-Food Economics and Natural Resources, Agri-Food Research and Technology Center of Aragón (CITA), Spain.

1. In the Finnish case, a sample of contact details was randomly selected from the total population of relevant businesses.
2. The Spanish sample was collected in the Aragón region. In the UK sample more than 80% are from Scotland. Despite these regional biases these samples are referred to as Spanish and UK samples.
3. The AMOS software package (version 18) was used with unbiased covariances as the input matrix. For missing values in the dataset, maximum likelihood estimation was conducted.
4. A two-level, six-item Relationship Sustainability construct, made up from two latent factors (Relationship Quality and Relationship Stability, both measured by three different items, as in Reynolds et al. 2009) was also tested. However, the final overall model fit was less satisfactory, it was thus decided to estimate the discussed SEMs based on the one-level, four-item Relationship Sustainability construct.
5. As shown by parameter mean-difference test which are not reported here.

References

Bollen, K. (1989). *Structural Equations with Latent Variables.* New York: John Wiley and Sons, Inc.

Cannon, J. and Homburg, C. (2001). Buyer–supplier relationships and customer firm costs. *Journal of Marketing* 65 (January): 29-43.

Cloud, J. (2007). Eating better than organic. *TIME Magazine*, March 2.

Donald, B. et al. (2010). Re-regionalizing the food system? *Cambridge Journal of Regions, Economy and Society* 3: 171-175.

Fischer, C. (2004). Managing international trade of food products—a survey of German and Australian companies. *Agribusiness: An International Journal* 20 (1): 61-80.

—. et al. (2007). Trust and economic relationships in selected European agri-food chains. *Food Economics* 4: 40-49.

—. and Hartmann, M. (eds.) (2010). *Agri-food Chain Relationships*. Wallingford, Oxford and Cambridge, Massachusetts: CAB International.

—. and Reynolds, R. (2010). Collaborative advantage, relational risks and sustainable relationships: a literature review and a definition. In Fischer, C. and Hartmann, M. (eds.), *Agri-food Chain Relationshipsi*, 74-87. Wallingford, Oxford and Cambridge, Massachusetts: CAB International.

—. et al. (2009). Factors influencing contractual choice and sustainable relationships in selected agri-food supply chains. *European Review of Agricultural Economics* 36 (4): 541-569.

—. et al. (2010). Determinants of sustainable agri-food chain relationships in Europe. In Fischer, C. and Hartmann, M. (eds.), *Agri-food Chain Relationships*, 119-134. Wallingford, Oxford and Cambridge, Massachusetts: CAB International.

Ganesan, S., Malter A. and Rindfleisch, A. (2005). Does distance still matter? Geographic proximity and new product development. *Journal of Marketing* 69 (October): 44-60.

Garbarino, E. and Johnson, M. (1999). The different roles of satisfaction, trust, and commitment in customer relationships. *Journal of Marketing* 63 (April): 70-87.

Ghemawat, P. (2001). Distance still matters: the hard reality of global expansion. *Harvard Business Review* (September): 138-147.

Hinrichs, C. (2003). The practice and politics of food system localization. *Journal of Rural Studies* 19: 33-45.

Kneafsey, M. (2010). The region in food–important or irrelevant? *Cambridge Journal of Regions, Economy and Society* 3: 177-190.

Martinez, S. et al. (2010). *Local Food Systems: Concepts, Impacts, and Issues*. Economic Research Report (ERR) 97, US Department of Agriculture, Economic Research Service, Washington, DC.

Reynolds, N., Fischer, C. and Hartmann, M. (2009). Determinants of sustainable business relationships in selected German agri-food chains, *British Food Journal* 111 (8): 776-793.

Sorenson, O. (2003). Social networks and industrial geography. *Journal of Evolutionary Economics* 13: 513-527.

CHAPTER THREE

RURAL DEVELOPMENT AND LOCAL AGRI-FOOD SYSTEMS: A NEW PARADIGM

EVY METTEPENNINGEN,
VALERIE VANDERMEULEN
AND GUIDO VAN HUYLENBROECK

Abstract

The rural development paradigm that emerged around the turn of the millennium represents a new paradigm for an agricultural development that tries to overcome the negative externalities related to intensification while emphasizing its broader role in the wider rural economy. In this chapter, we further develop this idea of a new development path for an agriculture which is not only embedded in the development of rural areas, but in our opinion plays a crucial role if we believe that rural development should be based on the specific characteristics and identity of rural areas. An important catalyst of this new paradigm are local agri-food systems. The chapter starts from recent socio-economic developments and challenges at the macro- and micro-level for the agricultural sector. It then describes the rise of the rural development paradigm, and tries to provide evidence for why this paradigm has not lived up to its expectations. Finally, the challenges which were identified and the incapability of the rural development paradigm to adequately address these challenges are combined in a proposal for a new paradigm. In this, regional identity, and local food systems in particular, serve as an engine for rural development.

1. Introduction

Since the second half of the 20th century, the agricultural sector in industrial countries has undergone substantial changes characterized by a constant decrease of labour, increasing farm size and a growing specialization of farm businesses (OECD 2002). An important driver for this structural change are agricultural policies, in particular in Europe the Common Agricultural Policy (CAP), first provided incentives for production increases and modernisation of farms and at a later stage has tried to mitigate the adverse effects of overproduction. The 1992 MacSharry reforms of the CAP, for example, reinforced the necessity for farm resizing because of decreasing per hectare margins (Moreno-Perez and Ortiz-Miranda 2008). Besides policies, technological developments (Flaten 2002) and rapid capital deepening[1] has pushed labour out of farming toward more labour intensive and better remunerating non-farm sectors (Dennis and Iscan 2009). Notwithstanding this general trend, structural changes in agriculture are subtle and spatially differentiated (Lobley and Potter 2004). While in some areas intensification of agriculture is dominating, in other areas agriculture is extensifying or diversifying, or new types of productivism linked to external capital are appearing. The rural development paradigm that emerged around the turn of the millennium provides a frame for agricultural development that tries to overcome the negative externalities related to intensification while emphasizing its broader role in the wider rural economy.

In this chapter, we further develop this idea of a new development path for an agriculture which is not only embedded in the development of rural areas, but in our opinion plays a crucial role if we believe that rural development should be based on the specific characteristics and identity of rural areas. An important catalyst of this new paradigm based on regional identity are local agri-food systems. In the following two sections of this chapter, recent socio-economic developments and challenges at the macro- and micro-level for the agricultural sector are described. Section 4 focuses on the rise of the rural development paradigm, and tries to provide evidence for why this paradigm has not lived up to its expectations. In Section 5, the challenges presented in Sections 2 and 3 and the inadequacy of the rural development paradigm in addressing these challenges are combined in a proposal for a new paradigm. In this, regional identity and local food systems in particular serve as an engine for rural development. The chapter concludes with a couple of recommendations on how to make the new paradigm work.

2. Macro-trends influencing agriculture

According to Porter (1990), important external factors influencing an industry will relate to: factor conditions, demand conditions, conditions linked to related and supporting industries, and policies. As the policy aspect will be discussed in detail in Section 4, this section will focus only on the first three aspects.

In the case of agricultural production, factor conditions will relate mainly to shifts in the land factor, which refers to all natural resources used in the production process, labour or capital. One of the major macro-trends within this category is the growing world population. A recent press release by the United Nations predicted an increase in the world population from the current 7 billion people to 9.3 billion by 2050 and 10.1 billion by the end of the century (United Nations 2011). This growing population will increasingly be located in urban areas, as there is a global trend towards urbanization. Urban areas are also expanding, in what is referred to as urban sprawl, as a result of several factors such as a change in living preferences, increased possibilities for mobility, and a lack of adequate spatial planning. One of the consequences of urban sprawl is increased land and soil consumption, which reduces the available space for agriculture (European Environment Agency 2006). Concerning rural areas, the rural population in Europe has been declining since the first half of the 20th century, and by 2050 it will have decreased from 199 million to 109 million people (United Nations 2009). The decrease in the agricultural labour force, as one important rural actor, is discussed more in detail in Section 3. Alongside population growth, the economy is constantly expanding as well, entailing a growing need for inputs in production processes. This however entails a decreasing stock of fossil inputs, such as crude oil. Hubbard peak oil theory predicted a fall in global oil production in the first decade of the 21st century (Bardi 2009). There is not only a problem with sources for production, but also a problem of assimilating the ecological consequences of economic productivity. An example is climate change, caused by an accumulating amount of greenhouse gasses, such as CO_2, in the atmosphere. According to the IPCC (Intergovernmental Panel on Climate Change), to limit the global increase in temperature to 2°C, the CO_2 concentration must be stabilized at 450 ppm. This requires an 85% reduction in global CO_2-emissions by 2050 compared to the level of 1990 (IPCC 2007). According to some authors, however, the 450 ppm level will not be sufficient to attain this goal (Allen et al. 2009; Meinshausen et al. 2009). Climate change is already proving a big challenge for farmers, who have to cope with greater variability and unpredictability of production (droughts, floods, movements of populations, etc.). Because climate change could only be mitigated by large-scale and costly effort, this challenge is not likely to become easier in the near future. Climate change results in an increased volatility of agricultural markets, currently being made worse by repercussions of the 2007 economic crisis.

In demand several macro-trends driven by consumer behaviour and social evolution can be distinguished. Firstly, population growth will result in an increasing demand for food and a need for further investment in agricultural production. Secondly, a change in income can lead to a convergence in demand for food. As income rises, calorie intake will increase, and will do so in more expensive forms. Typical examples as income level rises are a higher consumption of meat and dairy and a lower consumption of cereals. It is important to note that this increase in consumption is only in absolute terms, since the proportion of the budget spent on food is decreasing with increasing income (Engel's law) (Frazão et al. 2008). Changes in income, however, influence not only diet composition, but also preferences regarding quality and origin of food. An example is the growing interest in local food, although this is also a response to the trends towards standardization and industrialization in times of globalization mentioned above (Bowen and De Master 2011). The Slow Food movement, which originated in the mid-1980s in Italy, specifically promotes local food, traditional forms of cooking and eating food at a slow pace in a socially pleasant environment, in significant contrast to the fast food industry. Jones et al. (2003) however doubt that the movement will be strong enough to counter fast food or bring about major changes in the world's eating habits. Thirdly, agriculture is facing more and more societal pressure regarding environmental problems. Research by Givens and Jorgenson (2011) in 38 nations has indicated that environmental concern increases with economic development and growth in environmental degradation. Because of the high visibility of climate change, in atmospheric and water pollution, decreasing biodiversity and other areas, there is an increasing pressure on agriculture to improve its sustainability. As a result, policy instruments at European level such as the Water and Nitrate Directive try to limit the negative impact of agriculture on the environment. With increasing urbanization, the pressure on agriculture as a source of pollution is likely to grow, as people who have grown up in an urban area show higher levels of environmental concern (Czap and Czap 2010). As well as societal concern over the environmental effects of agriculture, there is also concern related to the use of new food technologies. Especially those technologies which are characterized as "bioactive" are often not trusted because of their unpredictable effects and because of ethical concerns (Frewer et al. 2011). One of these technologies is Genetically Modified Food products. According to Dannenberg (2009), the aversion to GM food is increasing sharply in Europe, increasing slightly in the US, and even decreasing in the rest of the world.

The macro-trends, linked to related and supporting industries (Porter 1990), concern shifts in the relationship between farmers and the agribusiness complex, as well as with farmers abroad, so that international trade is included. Globalisation is an important trend. Between 2000 and 2008, world exports increased by 50% and the volume of exports from developing countries nearly doubled (World Trade Organization 2011). This is leading to a larger variety in foods available from all over the world (especially for high-income groups), but at the same time to a convergence in sometimes unhealthy consumption habits. This is sometimes termed Coca-Colonization and particularly affects low-income groups (Trichopoulou et al. 2002; Rayner et al. 2006). It is linked to the rise of new diseases such as obesity and related cardiovascular conditions, although recent research for the US has shown that there is no clear link between socio-economic status and being overweight (Wang and Zhang 2006). Secondly, as a result of urbanization, there is an increasing distance between the site of production and the site of consumption. This leads to a higher involvement of the retail sector in the food chain (Traill 2006), which entails an increased need for standardization. As Murdoch et al. (2000, 119) state:

> Indeed, the requirement to abide by the guidelines associated with traceability and "supply chain management" often forces producers to replace their traditional production practices with those prescribed by the supermarke.

Moreover, the food system is controlled by a decreasing number of multinational companies, which can further decrease the available variety and limit consumers' choices (Schwartz and Lyson 2007). Thirdly, there is the decreasing stock of fossil fuels, mentioned above. A possible way to reshape the current economy so that it is better fit to deal with this challenge, would be to increase the use of biomass as a source for production of energy and raw materials. In 2010, about 5% of the economy in the European Union was biobased (Vandermeulen et al. 2010). This increasing demand for biomass for non-food applications is creating the opportunity for the agricultural sector to raise the value of its output, but at the same time is increasing land and food prices. The increasing use of land for the production of biofuels was one of the reasons for the sharp increase in food prices in the course of 2008.

All this is leading to major pressures on today's agriculture. Globalization in food production and pressure due to climate change are making food production less predictable and variable. In combination with greater variability in demand caused by recession and emerging non-food uses of agricultural produce, this is giving rise to greater market volatility. The highly unpredictable context in which agriculture operates is exacerbated by increased societal pressure on agriculture. These macro-trends have a direct influence on agriculture, and are leading to profound changes at the micro-level. In the next section, the focus is on one of the most important micro-trends influencing European agriculture, the de-agrarization of rural areas.

3. Micro-trends influencing agriculture: the de-agrarization of rural areas

Home to more than 56% of the population and covering more than 91% of the territory, rural areas play an important role in the European Union. Judging from European statistics on the EU-27, the role of agriculture in these rural areas is constantly declining. The most recent data available shows an 11.8% decrease in the agricultural labor force from 2003 to 2007 (Eurostat 2010). According to Copus et al. (2006), since 1990 there has been an almost universally downward trend of 2-3 % per year in agricultural employment in the EU-15. The share of agricultural employment in total rural employment is less than 10% for most rural regions in the EU-27 and for one third of the rural regions even less than 5%. However, in Central and Eastern European countries and the Southern part of the EU, the share of agriculture in total rural employment can be above 25% (Copus et al. 2006). In 2007, 75.5% of this agricultural labor in the EU-27 can be characterized as family labor, indicating the still predominant role of family farms. The decrease in the agricultural labor force goes together with a decrease in the number of holdings. From 2003 to 2007, there was a 6.5% decrease in total number of agricultural holdings (EU-27). The agricultural surface area utilized, however, stayed relatively stable with only a slight decrease of 0.5% from 2003 to 2007 (EU-27) (Eurostat 2010).

Not only is there a decreasing number of farmers in European rural areas, there is also an inflow of people from urban areas. According to Copus et al. (2006), the net-migration between 1995 and 2000 in significantly rural and predominantly rural areas in the EU-25 respectively amounted to 1.30 and 0.82%, whereas in significantly urban areas this was -0.13%. Research by van Dam et al. (2002) in the Netherlands also revealed a significant demand for rural living, which is linked to a very positive image of the countryside or the idea of a "rural idyll". Next to that, and even more importantly, rural and peri-urban areas are attracting more visitors from urban areas for recreation activities thanks to the increase in leisure time and increasing mobility. Rural areas are increasingly becoming a space for consumption by the urban population. This inflow of urban residents, together with the influence of the macro-economic trends described in the previous section is creating a different pattern of expectations towards the agricultural sector. Buijs et al. (2006, 386) describe this shift from functional towards hedonistic preferences for rural landscapes: "Where formerly the hiker hiked through the farm landscape, nowadays the farmer farms in the leisure landscape".

Literature does not provide a lot of information on what the public demands from agriculture (Hellerstein et al. 2002; Hall et al. 2004). Hellerstein et al. (2002) tried to investigate public requirements from agriculture indirectly by looking at the objectives specified in state legislation to protect farmland. They found that although environmental management can be carried out by other actors as well as farmers, legislation still ensures an important role for farmers. Hellerstein et al. (2002) state that this reveals an emotional attachment of the public to an agricultural base. Hall et al. (2004), although basing their conclusions on partial and statistically non-robust studies, also conclude that the public still values the farmers' role as providers of environmental goods and services and also cultural heritage.

More recently, Zasada (2011) presented a review of societal demands on agriculture in peri-urban areas. In line with the conclusions of Hellerstein et al. (2002) and Hall et al. (2004), Zasada (2011) noticed that farmers' contribution to the environment and the landscape is especially highly valued. However, there seems to be a higher recognition for the visual aspects of farmland than for its ecological value. Moreover, the preferences of the urban population do not match well with the trend towards scale-enlargement and specialization in the agricultural sector; the public prefer a landscape with small-scale and heterogeneous farm structures together with natural elements. The public tend to value peri-urban agriculture for delivering recreational and social services as well as its contribution to nature and the landscape. According to Pedroli et al. (2007, 434) urban dwellers have become increasingly disconnected from nature due to the "virtualities" of the city (like neon-lights, traffic, noise, etc.). Spending time in the countryside can relieve urban stress and help them to reconnect with their "own human essence". However, some landscapes have a bigger effect than others. It seems that in our globalized society people have an increased need for places with an identity they feel a bond with (Castells 1997). Pedroli et al. (2007) also acknowledge this role of rural landscapes as a provider of identity, and claim that local specificities of landscapes can better be accommodated by small-scale agriculture than by industrial farm types. Finally, Zasada (2011) mentions increasing demand by urban dwellers for regional food products. In Section 6, we will look in more detail at agri-food systems.

From this section we can conclude that the macro-trends described in Section 2 also lead to profound changes at the micro level, where we can identify an increased demand but also pressure from the urban population towards rural areas and the agricultural sector in particular. Some of these changes, sometimes accompanied by legislation, have led to a reorientation of the agricultural sector towards a more multifunctional way of farming embedded in the rural area as a whole. This rural development paradigm, part of a post-productivism movement in the agricultural sector, will be the focus of the next section.

4. The rise and fall of the rural development paradigm

After the Second World War, the main objective of the CAP in Europe was to ensure adequate supply of food for the population, increase the agricultural income and the competitiveness of the sector. Although the policy instruments had their intended effect, they also led to agricultural surpluses, negative environmental externalities and aroused resistance within the World Trade Organization (WTO). The MacSharry reforms of the CAP in 1992 tried remove these negative effects by reducing market support and replacing it partly by compensation payments. It also introduced the possibility for the EU member states to give support for agri-environmental measures and measures aimed at structural change, such as early retirement schemes. These reforms, however, were still unable to solve the problems faced by rural areas at that time. As well as the changing role of agriculture in society from production to consumption, as described in the previous section, rural areas also had their specific concerns as expressed in the 1996 Cork Declaration. The Cork Declaration demanded a fairer balance in public investment between rural and urban areas, and a multi-sectoral approach to rural development with a territorial dimension (European Commission 1996). Farmers also themselves expressed doubts about the productivism paradigm. Research by Ettema (1994, cited in van der Ploeg et al. 2000) in the Netherlands revealed that only 15 to 20% of the farm entrepreneurs supported the continuous scale enlargement in the sector and the resulting rural exodus stimulated by the CAP. Van der Ploeg and Roep (2003) contest the image of the farmer as merely a profit seeker, and stress that farmers may also have non-commercial motivations, such as the maintenance of the cultural heritage, spreading risks, family life, residential and lifestyle preferences. With the decreasing relative importance of the farming population in rural areas, there was also the insight that other rural actors, such as hobby farmers and rural dwellers, could play a more important role in managing nature and the landscape (van der Ploeg et al. 2000), improving social cohesion, quality of life and providing rural employment (Renting et al. 2009).

The resulting changes in the CAP came with the Agenda 2000 reforms. With Agenda 2000, the importance of rural areas was officially acknowledged by the EU, who introduced a second pillar devoted to rural development into the CAP.[2] This was accompanied by the request for member states to develop their own rural development program. These reforms marked the change from the productivism paradigm to the rural development paradigm, embedding agriculture in the wider rural economy. In the rural development paradigm, agriculture was considered to be inherently multifunctional: as well as producing food and fibre, agriculture takes care of nature and the environment and can play a key role in increasing the welfare of rural and urban populations. The Agenda 2000 policy also made a modest start to enlarging support for CAP beyond the agricultural sector by providing funds for other rural sectors through Axis 3 and LEADER. According to Marsden and Sonnino (2008), these changes in the CAP indicate a shift from a sectoral to a regional or territorial perspective on rural areas, in which agriculture is integrated into rural development. Van der Ploeg et al. (2000) believe that a key characteristic of this rural development paradigm is the pursuit of synergy and cohesion between farming styles, different ecosystems, localized food chains and social movements. The aim of the rural development paradigm is to create win-win situations for the rural area as a whole.

With the end of the second programming period (2007-2013) of rural development programs slowly approaching, and discussion on the next period now underway, an increasing number of evaluations of the effects of European rural development policies are appearing. First of all it seems that the subsidies have only a modest impact on the economic performance of rural areas (Diakosavvas 2006). The effect of the second axis' agri-environmental payments is also under discussion. Although it may be too soon to see the overall effects, many studies are critical of the schemes' ability to create the intended environmental goods and services (Kleijn and Sutherland 2003; Kleijn et al. 2004; Berger et al. 2006; Kleijn et al. 2006; Matzdorf et al. 2008; Ohl et al. 2008). The overall critique is that the policy instruments in the second axis are too rough and not able to fine-tune policy support to the diversity of interactions between agriculture and the environment (Petersen 2010). Since professional farmers are the only beneficiaries for some of these measures, some pluri-active or diversifying farms which do not officially qualify as agricultural holdings cannot receive subsidies (Laurent et al. 2002), which is a potential weakness. Axis 3 and LEADER, aiming at the wider rural population, seem to be powerful tools to create change in rural areas. However, increased criticism of the 3rd and 4th axis is being heard, as the changes implemented are considered to be too fragmentized. Moreover, projects are often set up only because money is available and not because of a real need within the rural community (Oedl-Wieser et al. 2010). Contrary to its intentions, the LEADER program may also lead to social exclusion, being primarily used by social elites to restore legitimacy or for clientelism. Another problem linked to LEADER, although also applicable to other elements of rural development policy, is that trying to reach rural development through projects with competitive bidding might compromise the transparency of the policy and democratic accountability (Shucksmith 2000).

The literature finally also provides some general critiques on the rural development paradigm and how it is translated into the second pillar of the CAP. On the basis of observations in the UK, Marsden and Sonnino (2008) conclude that there are substantial contradictions between agricultural policies at the national level. On the one hand, the agri-food policy stimulates a rational and retailer-led food supply system with an industrial farm model, while on the other hand the rural development policy tries to decentralize and diversify the agricultural sector. The same authors also criticize the lack of innovation in rural development policies, which should create opportunities for creative forms of supply and demand management in the agri-food sector, which are more spatially embedded (Marsden and Sonnino 2008). Within the second pillar of the CAP, Sumelius and Bäckman (2008) believe policy makers have made too much use of the "one instrument per objective to be achieved" principle, resulting in policies for multifunctional agriculture which are not well integrated. This not only influences the effectiveness of the instruments in achieving the desired changes in rural areas, but also the costs involved. One policy measure designed to achieve several objectives at once probably involves less transaction costs than several separate instruments (Vatn 2001, cited in Renting et al. 2009). An important aspect of these transaction costs, which is often neglected, is the cost of setting up and maintaining the different institutional structures responsible for implementing the different policy measures (Knickel and Kröger 2008). Then there is also the aspect of evaluation of rural development policies, which, according to Knickel and Kröger (2008), focuses too much on environmental aspects, and neglects important aspects such as quality of life, energy use, social inclusion and the link of the agricultural sector with activities such as leisure and recreation. Another problem in evaluation is that it can be difficult to reconcile rural development policies with the aim of giving more power to the people. From 1989 onwards, support from the EU structural funds to certain rural areas in Europe was linked to a strategy for those regions, developed in cooperation with the local people. However, there is increasing complaint from local people that the parameters set by the state are too tight and that there is too much control (Shucksmith 2010). On the other hand, a certain amount of control is necessary to avoid misuse of funds. According to Shucksmith (2010) there is a need to move away from the integrated model of rural development to a disintegrated one, which means moving from a "government" to a "governance" approach. Governance means not only sponsoring local projects, but delivering them through partnerships between the government and local stakeholders

All these critiques indicate that the current rural development paradigm is not yet fully adequate to deal with the specific challenges rural areas are facing nowadays. Focussing on the agricultural sector, Renting et al. (2009) believe that there are still too many weak forms of multifunctionality, and that the concept and its claim of providing positive externalities is often merely (mis)used to legitimize subsidies. For these authors, a strong multifunctionality of agriculture would entail incorporating the sector into an integrated and territory-based rural development approach (Renting et al. 2009).

On the basis of criticisms on the current rural development paradigm and the challenges agriculture and rural areas are facing nowadays, the next section argues that a new rural development paradigm is needed which puts regions and their specificities at the centre. In this paradigm, agriculture both contributes to and profits from a regional dynamic in which economic and ecological cycles within the region are closed.

5. Towards a new paradigm: regional identity and local food systems as an engine for rural development

The previous section showed that the current rural development paradigm aiming to fit agriculture within rural areas is facing several problems. Most important is that its effects are still very marginal, that there is too much involvement of central governments and that its level of spatial embeddedness is not sufficient. This last aspect is especially disturbing, given the fact that regions and their identity are becoming more important economically and otherwise. Sections 2 and 3 of this chapter revealed an increased interest of people in regions and regional foods as a result of or reaction to macro- and micro-economic trends like increasing globalization, ecological concerns, health issues and lifestyle preferences. Based on these concerns, a new rural development paradigm in which regions and their specificities play a central role which appears very promising. A central aspect of this new rural development paradigm is regional identity. Regional identity involves the characteristic elements of an area which distinguish it from other areas, and is closely linked to regional or local history, traditions, culture and folklore. However, regional identity is more than the existence of such characteristic elements, since it is not only based upon the meaning attached by local people or visitors to these elements but also on the way people develop an attachment to the region (Kruit et al. 2004). In other words, regional identity is a social construction implying that regions can also have more than one identity depending on the different groups of people living or visiting this region (Paasi 2003).

It is clear that because agriculture uses and manages a large share of the land in rural areas, it has a significant influence on the identity of a region. Traditional agricultural practices and typical food products are important aspects of regional identity (MacDonald et al. 2000; Dramstad et al. 2001; Calvo-Iglesias et al. 2006; Daugstad et al. 2006). Figure 3-1 schematically presents a new rural development paradigm which is based on the economic and ecological value of regional identity and local food systems. In this paradigm, we start from the observation that agriculture has an important influence on the identity of a region, for example through local food systems. Since regions are becoming more important for people, stimulating regional identity coincides with meeting societal demands. As people show an increased preference for authentic (local) products and places in their consumption patterns (Sims 2009), this regional identity can create economic value and possibilities for regional economic development. Agriculture in turn can also benefit from this increased competitiveness, e.g., as a supplier of local food products, and also as provider of tourism and recreation facilities. In this way, the blue circle in Figure 3-1 is closed.

Next to the blue circle, the figure also shows a green circle. Not only does this new paradigm meet the societal demands related to increased globalization and lifestyle preferences, it also meets the ecological concerns of the population as a more localized style of consumption entails an improvement in environment quality. This reality is understood by a new generation of environmentalist organizations, striving for a more localized style of living as an answer to problems like climate change and peak oil production (Hopkins 2008). Consuming local products and organizing leisure activities closer to home are in fact ways of reducing ones carbon footprint. A diversified agricultural sector can respond to both of these demands. Moreover, the production of biomass, and in particular the possibility of second generation biofuels, offers another opportunity for agriculture to satisfy the needs of the new ecologically conscious generation and to close local production cycles. This would be in line with the Cradle to Cradle principle, which argues for a complete closure of local production cycles with all waste of local production processes being re-used (Braungart et al. 2007).

Figure 3-1 A new rural development paradigm based on regional identity and local food systems

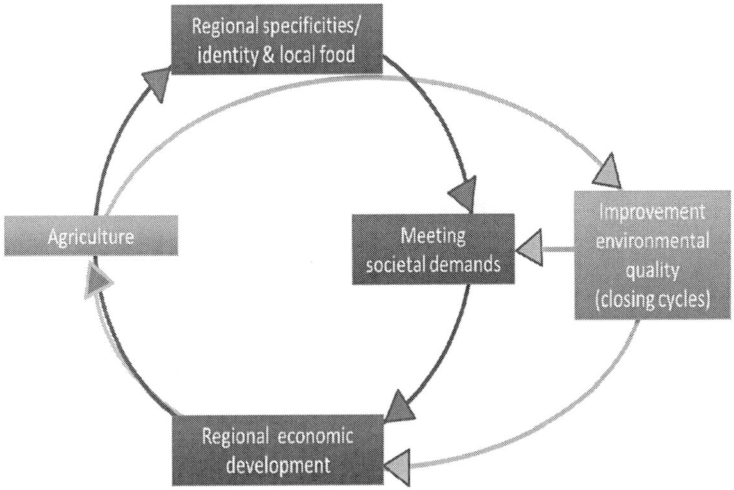

Different aspects within this new paradigm can strengthen each other. A strong regional identity with characteristic landscapes, traditions and cultural heritage can attract visitors to an area (Torres 2002). A possible strategy for a region to profit from this tourism is to provide locally-sourced products, for example through farmers' markets (Tregear et al. 2007). An increased awareness of health and environmental problems, related to food scares and the knowledge that food has the highest share in the household's impact on climate change, may increase the demand for these local produced food products. Next to satisfying consumer demand, this local food system can also be a response to the growing concerns of farmers, mentioned in the previous section, who expect more from farming than merely making profit. As such, these local food initiatives can form a response to economic, environmental and social challenges farmers are facing today (Sims 2010).

Notwithstanding its potential, this new paradigm is also vulnerable to a number of challenges. First of all, there are challenges related to the farmers. The modernization of agriculture has put pressure on the traditional role of farmers as contributors to regional identity and traditions. Regional development organizations are only slowly recognizing the importance of involving farmers in regional development and branding processes to avoid negative effects (Mettepenningen et al. 2010). The incentive for farmers to engage in rural development processes is still low and is expected to further decrease in the future with the increasing possibilities of using biomass in industrial processes (Dale 2007). Farmers interested in producing local food products also face legal and regulatory constraints such as more severe requirements regarding food quality and safety. These demands, originally coming from consumers, have become stricter with an increasing involvement of retailers and food companies in the distribution of local food. This may lead to the paradoxical situation that farmers applying more traditional farming practices but who don't possess the necessary capital, skills or knowledge may face exclusion from this traditional local food systems (Challies and Murray 2011). Another possible threat is that people express the desire to buy products from local farmers, but are not willing to pay for it either in monetary terms or in effort to find the outlets where these products are sold (Smithers et al. 2008). Research by Chambers et al. (2011) revealed that in general consumers often do not understand the challenges linked to sustaining natural resources. According to these authors, bioregional marketing which connects consumers to the agricultural landscape may help in overcoming this problem and improve development prospects for local food systems.

Secondly, local food systems can, paradoxically, pose a threat to locality and local identity. One of the threats is that developing traditional food products entails new legislation for protection of producers and typical production modes. But as mentioned above, not all farmers are able to comply with these laws. As a result, there may be involvement by external actors who patent these products, get access to the niche markets, but do not value the traditions as locals do. Tradition is in this way "commodified". The exclusion of local producers can lead to a change in the power relations in the traditional food supply chain, creating income inequality and social exclusion. The increased need for monitoring and easily measurable standards deriving from a local label may lead to standardization and homogenization of the products with a potential loss of specific characteristics. On the other hand, preserving tradition can lead to a kind of "disneyfication" of agriculture and rural areas which is also not a desirable effect. In local food systems, modern practices increasing efficiency have a place. In setting up codes and standards, official institutions even though designed to protect local food products might in fact change the cultures they were designed to protect. In some cases however, e.g., Comté cheese production in France, this institutionalization can work and local farmers reap the benefits (Bowen and De Master 2011).

A third challenge is that the value and benefits of local food systems for the environment and health are not always evident. Using the concept of food miles is a good way to catch consumer attention for local food, although sustainability is not solely determined by distance. Greenhouse gas emissions can also be linked to production, as can land management, farm inputs, the skills of the farmer and storage facilities. Edwards-Jones (2010) gives the example that consuming New Zealand apples shipped to the UK in spring may be more environmentally friendly than consuming UK apples from the last autumn stored in energy consuming refrigerators. Linked to this is the health debate. Storing local vegetables or fruits to be able to consume them out of season leads to a loss of nutritional value. So from a health perspective, it might be better to consume fruits and vegetables which are imported from a region where they are in season (Edwards-Jones 2010).

6. How to make the new paradigm work?

In order tor the new rural development paradigm based on regional identity and local food systems to work, some critical preconditions need to be fulfilled. First of all, it is important to start from a well developed strategic plan or vision for the region, based on its specific strengths and weaknesses. Not every region, for example, will benefit from the development of local food chains. In creating this vision, there is a need for a broad involvement of local stakeholders and strong political support (Mantino 2010). This approach also calls for a decentralization of governance structures. However, making such structures function requires an increase and investment in local social and human capital and this process may require time. So public authorities need to take responsibility and participate at the local level in the first stages of this process, without compromising the need for flexibility. There need to be found new institutional arrangements which positively influence the supply of private and public goods, provide ways to cope with market failure, save on transaction costs and support the building of relevant social and learning networks (Van Huylenbroeck et al. 2007).

Notes

1. Or increase in capital intensity.
2. The second pillar contains 4 Axes: Axis 1 on improving the competitiveness of the agricultural and forestry sectors; Axis 2 on improving the environment and countryside; Axis 3 on improving the quality of life in rural areas and encouraging diversification of the rural economy and Axis 4 (LEADER) on building local capacity for employment and diversification.

References

Allen, M.R. et al. (2009). Warming caused by cumulative carbon emissions towards the trillionth tonne. *Nature* 458 (7242): 1163-1166.

Bardi, U. (2009). Peak oil: The four stages of a new idea. *Energy* 34 (3): 323-326.

Berger, G. et al. (2006). The greening of the European common agricultural policy by linking the European-wide obligation of set-aside with voluntary agri-environmental measures on a regional scale. *Environmental Science & Policy* 9: 509-524.

Bowen, S. and De Master, K. (2011). New rural livelihoods or museums of production? Quality food initiatives in practice. *Journal of Rural Studies* 27 (1): 73-82.

Braungart, M. et al. (2007). Cradle-to-cradle design: creating healthy emissions—a strategy for eco-effective product and system design. *Journal of Cleaner Production* 15 (13-14): 1337-1348.

Buijs, A.E. et al. (2006). From hiking through farmland to farming in a leisure landscape: changing social perceptions of the European landscape. *Landscape Ecology* 21: 375-389.

Calvo-Iglesias, M.S. et al. (2006). Exploring farmer's knowledge as a source of information on past and present cultural landscapes—A case study from NW Spain. *Landscape and Urban Planning* 78 (4): 334-343.

Castells, M. (1997). *The Power of Identity. The Information Age: Economy, Society and Culture*. Malden, USA, Blackwell Publishing.

Challies, E.R.T. and Murray, W.E. (2011). The Interaction of Global Value Chains and Rural Livelihoods: The Case of Smallholder Raspberry Growers in Chile. *Journal of Agrarian Change* 11 (1): 29-59.

Chambers, K.J. et al. (2011). What does "local" mean in the grocery store? Multiplicity in food retailers' perspectives on sourcing and marketing local foods. *Renewable Agriculture and Food Systems* 26 (1): 46-59.

Copus, A. et al. (2006). Study on Employment in Rural Areas. Final Deliverable. A study Commissioned by: European Commission, Directorate General for Agriculture. Unit F.3. Consistency of Rural Development. Contract No. 30-CE-0009640/00-32.

Czap, N.V. and Czap, H.J. (2010). An experimental investigation of revealed environmental concern. *Ecological Economics* 69 (10): 2033-2041.

Dale, B.E. (2007). Why this new journal? *Biofuels Bioproducts & Biorefining-Biofpr* 1 (1): 1-2.

Dannenberg, A. (2009). The dispersion and development of consumer preferences for genetically modified food—A meta-analysis. *Ecological Economics* 68 (8-9): 2182-2192.

Daugstad, K. et al. (2006). Agriculture as an upholder of cultural heritage? Conceptualizations and value judgements—A Norwegian perspective in international context. *Journal of Rural Studies* 22 (1): 67-81.

Dennis, B.N. and Iscan, T.B. (2009). Engel versus Baumol: Accounting for structural change using two centuries of US data. *Explorations in Economic History* 46 (2): 186-202.

Diakosavvas, D. (ed.) (2006). *Coherence of Agricultural and Rural Development Policies. The Development Dimension*. Paris, OECD.

Dramstad, W.E. et al. (2001). Integrating landscape-based values—Norwegian monitoring of agricultural landscapes. *Landscape and Urban Planning* 57 (3): 257-268.

Edwards-Jones, G. (2010). Does eating local food reduce the environmental impact of food production and enhance consumer health? *Proceedings of the Nutrition Society* 69 (4): 582-591.
European Commission (1996). The Cork Declaration—A living countryside. The European Conference on Rural Development.
European Environment Agency (2006). Urban sprawl in Europe. The ignored challenge. No 10/2006: 60.
Eurostat (2010) Agricultural statistics. Main results 2008-09. 2010 Edition. Luxembourg, Publications Office of the European Union.
Flaten, O. (2002). Alternative rates of structural change in Norwegian dairy farming: impacts on costs of production and rural employment. *Journal of Rural Studies* 18 (4): 429-441.
Frazão, E. et al. (2008). Converging Patterns in Global Food Consumption and Food Delivery Systems. *Amber Waves* 6 (1): 22-29.
Frewer, L.J. et al. (2011). Consumer response to novel agri-food technologies: Implications for predicting consumer acceptance of emerging food technologies. *Trends in Food Science & Technology* 22 (8): 442-456.
Givens, J.E. and Jorgenson, A.K. (2011). The Effects of Affluence, Economic Development, and Environmental Degradation on Environmental Concern: A Multilevel Analysis. *Organization & Environment* 24 (1): 74-91.
Hall, C. et al. (2004). What does the public want from agriculture and the countryside? A review of evidence and methods. *Journal of Rural Studies* 20: 211-225.
Hellerstein, D. et al. (2002). *Farmland protection and the role of public preferences for rural amenities report 815*, USDA Economic Research Service.
Hopkins, R. (2008). *The transition handbook. From oil dependency to local resilience*. White River Junction, Chelsea Green Publishing Company.
Intergovernmental Panel on Climate Change (2007). Climate Change 2007: Synthesis Report—Summary for Policymakers: 22.
Jones, P. et al. (2003). Return to traditional values? A case study of Slow Food. *British Food Journal* 105 (4/5): 297-304.
Kleijn, D. et al. (2006). Mixed biodiversity benefits of agri-environment schemes in five European countries. *Ecology Letters* 9 (3): 243-254.
—. et al. (2004). Ecological effectiveness of agri-environment schemes in different agricultural landscapes in the Netherlands. *Conservation Biology* 18 (3): 775-786.

Kleijn, D. and Sutherland, W.J. (2003). How effective are European agri-environment schemes in conserving and promoting biodiversity? *Journal of Applied Ecology* 40 (6): 947-969.

Knickel, K. and Kröger, M. (2008). Evaluation of policies in terms of the multifunctionality of agriculture: More integrative conceptual and analytical frameworks needed. *International Journal of Agricultural Resources, Governance and Ecology* 7: 399-418.

Kruit, J. et al. (2004). Regionale identiteit van natuur en landschap; Een verkenning van een containerbegrip en de bruikbaarheid als sturingsinstrument, Alterra report: 100.

Laurent, C. et al. (2002). Multifonctionnalité et égibilité aux aides PAC dans l'UE. *Économie rurale* 268-269: 144-158.

Lobley, M. and Potter, C. (2004). Agricultural change and restructuring: recent evidence from a survey of agricultural households in England. *Journal of Rural Studies* 20 (4): 499-510.

MacDonald, D. et al. (2000). Agricultural abandonment in mountain areas of Europe: Environmental consequences and policy response. *Journal of Environmental Management* 59 (1): 47-69.

Mantino, F. (2010). Understanding delivery mechanisms in EU rural development policies: an institutional approach. 118th EAAE Seminar: *Rural development: governance, policy design and delivery*. Ljubljana, Slovenia.

Marsden, T. and Sonnino, R. (2008). Rural development and the regional state: Denying multifunctional agriculture in the UK. *Journal of Rural Studies* 24 (4): 422-431.

Matzdorf, B. et al. (2008). Developing biodiversity indicator to design efficient agri-environmental schemes for extensively used grassland. *Ecological Indicators* 8 (3): 256-269.

Meinshausen, M. et al. (2009). Greenhouse-gas emission targets for limiting global warming to 2 degrees C. *Nature* 458 (7242): 1158-U1196.

Mettepenningen, E. et al. (2010). Perspectives for landscape management by farmers in the framework of regional branding processes. 9th European IFSA Symposium: *Building sustainable rural futures. The added value of systems approaches in times of change and uncertainty*. Wein.

Moreno-Perez, O.M. and Ortiz-Miranda, D. (2008). Understanding structural adjustment in Spanish arable crop farms: policies, technology and multifunctionality. *Spanish Journal of Agricultural Research* 6 (2): 153-165.

Murdoch, J. et al. (2000). Quality, nature, and embeddedness: Some theoretical considerations in the context of the food sector. *Economic Geography* 76 (2): 107-125.

OECD (2002). *Agricultural policies in OECD countries: monitoring and evaluation*. Paris, Organisation for Economic Cooperation and Development.

Oedl-Wieser, T. et al. (2010). The LEADER programme 2007-2013: refocusing towards agriculture-oriented activities as a consequence of "mainstreaming" innovative rural action in Austria. 118th EAAE Seminar: *Rural development: governance, policy design and delivery*. Ljubljana, Slovenia.

Ohl, C. et al. (2008). Compensation payments for habitat heterogeneity: Existence, efficiency, and fairness considerations. *Ecological Economics* 67 (2): 162-174.

Paasi, A. (2003). Region and place: regional identity in question. *Progress in Human Geography* 27 (4): 475-485.

Pedroli, G.B.M. et al. (2007). Values of rural landscapes in Europe: inspiration or by-product? *NJAS Wageningen Journal of Life Sciences* 54: 431-447.

Petersen, J.-E. (2010). Rural development policy as a tool for provision of ecosystem services and environmental public goods. 118th EAAE Seminar: *Rural development: governance, policy design and delivery*. Ljubljana, Slovenia.

Porter, M. (1990). *The competitive advantage of nations*. New York, Free Press.

Rayner, G. et al. (2006). Trade liberalization and the diet transition: a public health response. *Health Promotion International* 21: 67-74.

Renting, H. et al. (2009). Exploring multifunctional agriculture. A review of conceptual approaches and prospects for an integrative transitional framework. *Journal of Environmental Management* 90: 112-123.

Schwartz, R.A. and Lyson, T.A. (2007). Retail relations: an interlocking directorate analysis of food retailing corporations in the United States. *Agriculture and Human Values* 24 (4): 489-498.

Shucksmith, M. (2000). Endogenous development, social capital and social inclusion: Perspectives from LEADER in the UK. *Sociologia Ruralis* 40 (2): 208-218.

—. (2010). Disintegrated Rural Development? Neo-endogenous Rural Development, Planning and Place-Shaping in Diffused Power Contexts. *Sociologia Ruralis* 50 (1): 1-14.

Sims, R. (2009). Food, place and authenticity: local food and the sustainable tourism experience. *Journal of Sustainable Tourism* 17 (3): 321-336.

—. (2010). Putting place on the menu: The negotiation of locality in UK food tourism, from production to consumption. *Journal of Rural Studies* 26 (2): 105-115.

Smithers, J. et al. (2008). Unpacking the terms of engagement with local food at the Farmers' Market: Insights from Ontario. *Journal of Rural Studies* 24(3): 337-350.

Sumelius, J. and Bäckman, S. (2008). Review of studies on the establishment and management of policies for multifunctionality. *International Journal of Agricultural Resources, Governance and Ecology* 7: 386-398.

Torres, R. (2002). Towards a better understanding of tourism and agriculture linkages in the Yucatan: Tourist food consumption and preferences. *Tourism Geographies* 4: 282-307.

Traill, W.B. (2006). The rapid rise of supermarkets? *Development Policy Review* 24 (2): 163-174.

Tregear, A. et al. (2007). Regional foods and rural development: The role of product qualification. *Journal of Rural Studies* 23 (1): 12-22.

Trichopoulou, A. et al. (2002). Disparities in food habits across Europe. *Proceedings of the Nutrition Society* 61 (4): 553-558.

United Nations (2009). *2009 Revision of World Urbanization Prospects*. New York: United Nations.

United Nations (2011). *World Population Prospects, the 2010 Revision*. Press Release 3 May 2011. New York: United Nations.

Van Dam, F. et al. (2002). Lay discourses of the rural and stated and revealed preferences for rural living. Some evidence of the existence of a rural idyll in the Netherlands. *Journal of Rural Studies* 18 (4): 461-476.

Van der Ploeg, J.D. et al. (2000). Rural development: From practices and policies towards theory. *Sociologia Ruralis* 40 (4): 391-408.

Van der Ploeg, J.D. and Roep, D. (2003). Multifunctionality and rural development: the actual situation in Europe. In G. van Huylenbroeck and G. Durand (eds.), *Multifunctional Agriculture. A New Paradigm for European Agriculture and Rural Development*, 37-54. Aldershot, Hampshire, England: Ashgate.

Van Huylenbroeck, G. et al. (2007). Multifunctionality of Agriculture: A Review of Definitions, Evidence and Instruments. *Living Reviews in Landscape Research* 1 (3).

Vandermeulen, V. et al. (2010). *Hoe biobased is de Vlaamse economie? Departement Landbouw en Visserij—Afdeling Monitoring en Studie*. Brussels, UGent: 132.

Wang, Y. and Zhang, Q. (2006). Are American children and adolescents of low socioeconomic status at increased risk of obesity? Changes in the association between overweight and family income between 1971 and 2002. *American Journal of Clinical Nutrition* 84 (4): 707-716.

World Trade Organization. (2011). Millenium Development Goals. Trade and Development. Retrieved May, 5th 2011, from http://www.wto.org/english/thewto_e/coher_e/mdg_e/development_e.htm.

Zasada, I (2011). Multifunctional peri-urban agriculture—A review of societal demands and the provision of goods and services by farming. *Land Use Policy* 28 (4): 639-648.

CHAPTER FOUR

LOCAL FOOD QUALITY AND LOCAL RESOURCES

GIOVANNI BELLETTI, FRANÇOIS CASABIANCA AND ANDREA MARESCOTTI

Abstract

In recent times, the concept of "local food" has come to attention in academic and political discourse about food, usually closely linked to the growing debate on Alternative Agro-Food Networks. However, the meaning of "local food" it is not yet consolidated and two main meanings can be identified. The first refers to the relationship between consumer and producer: local food is food produced by firms located in places close to the final consumer. The second meaning is focused on the product and the resources used in its production process. Both these (non opposing) meanings often share "alternativeness" in the vision of food production and consumption, contrasting the dominant model of the globalised mass food chains. Both definitions use insights from new theoretical approaches to resources and space to emphasize the role of actors and the relevance of organizational proximity and local institutions in enabling the deployment of opportunities offered by local specific resources to local competitiveness.

This chapter explores the relationship between local food quality and local resources, starting from a discussion about the concept of local resources in economics, and exploring the different meanings of local food quality in the current debate about food.

1. Introduction, aims and methodology

In recent times, the concept of "local food" has become one of the most popular both in academic and political discourse about food, and is usually closely linked to the growing and more general debate on Alternative Agro-Food Networks.[1] However, the meaning of the expression "local food" it is not yet consolidated (Brunori 2007).

In fact, two main meanings of "local food" can be identified. The first refers to the relationship between consumer and producer: local food is food produced by firms located in places close to the final consumer. Food is therefore procured essentially by local sources through a wide set of marketing channels, such as direct sales, farmers' markets, community-supported agriculture, solidarity purchasing groups, public procurement, and also through modern channels (Dunne et al. 2010). In this definition, the focus is placed mainly on the physical proximity between producers and consumers, which is as short as possible (Mathjis et al. 2006).

The second meaning is focused rather on the product and the resources used in its production process. Here, local food is conceived as food with strong roots in a specific geographical place which gives the product its identity. The focus here is rather on the nature and specificity of the resources used in the production process.

These meanings do not conflict, and often share 'alternativeness' in the vision of food production and consumption, contrasting the dominant model of the globalised mass food chains (Fonte and Papadopoulos 2010).

Since the 1990s, the link between local food and local resources has emerged as critical issue in the debate on economic, social and environmental sustainability of food systems, although to date there has been little theoretical and empirical analysis. Local food is often advocated in literature as a means for reducing negative environmental impacts in the perspective of food miles (Pretty et al. 2005; Coley et al. 2009), but also considering origin products and their environmental links (Larson 2007; Thévenod-Mottet 2010). Some authors underline the role local food can play for keeping traditional production systems alive, especially those systems based on small and medium enterprises, and located in marginal areas (Bérard and Marchenay 2004; Barham and Sylvander 2011).

However, the need for a more critical attitude has been proposed by some authors (see, for example, Harris 2010), in order to take account of ambivalent aspects of the role of place in alternative food networks, and by extension the link between local food quality and local resources.

From a methodological point of view, it is necessary to develop interpretation frameworks and empirical analysis on specific aspects in order to accumulate experience and avoid harmful generalizations. This is the path that this chapter intends to pursue, with specific reference to the issue of animal agro-biodiversity. It aims at exploring the relationship between local food quality and local resources, and is structured in three parts.

The first part discusses the concept of local resource in economics and explores the different meanings of local food quality in the current debate about food. The two above mentioned dimensions of local food quality emerge: the first related to the physical proximity of producer and consumer ("local food" quality), the second focused on the way local specific resources are incorporated into (and determine the quality of) the product by means of the territorial anchorage of the production process ("local quality" of food). Both views use insights from new theoretical approaches to resources and space to discuss the role of actors and the relevance of organisational proximity and local institutions in deploying opportunities offered by local specific resources in competitiveness. Quality characteristics related to "local" allow activation, preservation and evolution of local resources, but failures in resources and markets can affect the way local resources are managed and renewed.

The second part is focused on Origin products, and presents an ideal typical "virtuous quality circle" linking local resources, products and markets in a self-sustaining relationship capable of renewing local resources. A key role is played by Geographical Indications, the potentialities of which are explored with reference to the European Union protection system.

In the third part empirical evidence on animal production will help to identify theoretical issues relevant for scientific debate and the development of place-based supporting public policies and private, collective enhancement strategies.

2. Food and resources: what is the meaning of "local"?

2.1 Local resources: a missing concept?

The concept of local resource is often absent in standard economics, or reduced to simple site specificity, or a "location" of a generic (standard) resource. Specific qualities of resources apart from the location are often not taken into account. Localised resources are more and more exploited on the basis of generic knowledge. Economic progress is accomplished by a process of replacement of "local specific" by "standardised resources" in production processes. In the case of food production, this is particularly evident for genetic vegetal and animal resources (OECD 2008), as well as technologies and know-how.

Recent theoretical and empirical approaches however are leading towards a re-consideration of the role of "place" and resources specific to a well defined geographical space.

> In the globalisation process of the economy, local factors and local specificities are fundamental elements upon which the competitiveness of countries [and regions] depend and therefore represent important areas where practitioners and policy makers require a sophisticated and advanced toolbox to intervene (Capello 2009).

New theories and approaches have re-evaluated the role of local resources (e.g., *Marshallian districts* in Italy, *Systèmes localisés* and *Milieux innovateurs* in France) (Pecqueur 2006; Requier-Desjardins 2009). Production processes are not only considered as place-based in a world of homogeneous resources, but also "locally situated", meaning that places are characterised by resources recognised as specific and able to influence the output of the production process, not as much as for their production cost, but rather for the quality of goods produced.

At the same time, these approaches widely recognise the complementarity between material resources and local knowledge, skills, relational abilities and cultural values orienting human attitudes specific to different places because they are built over time and not easily transferable and applied to the same processes in other places. Indeed, only these intangible factors (which are resources, too) can transform local resources into assets capable of creating value on the market. Some authors refer to this process of transformation of resources into assets, and of the territorial capital into heritage (patrimony), as "territorial anchorage" of resources and goods (Requier-Desjardins 2009). Territorial anchorage can be supported by specific policies, managed mainly at local level.

It is the complex net of relationships between resources and actors that matters: each territory has to be seen as a bundle of relations between a place/space and a population of actors, and their boundaries are necessarily fuzzy given that each actor keeps up relationships on different spatial scales. A territory is not a closed entity, but a "localised thickening" of relations (Becattini 1989).

Only organisational proximity can transform geographical proximity into a success factor. As underlined by institutionalist approaches (Hodgson 1998), local institutions and local rules (both formal and tacit) allow a better coordination between local actors and act as interface between "local resources" and "actors".

To summarize, location advantages for firms derive not only from spatial location and physical proximity, but mainly from relational and institutional proximity. The quality of local resources and the way they are used can therefore affect the quality of the good produced in each specific place. Other relevant factors influencing the re-emergence of "place" in a wider sense are the new consumer environmental and social concerns, which make necessary a reconsideration of the role of local resources in a wider perspective. Much literature emphasises and analyses the "dark side" of a-spatialised (globalised) production methods. With specific reference to food production there is concern over the environment (pollution, loss of biodiversity, soil erosion…), social problems (exploitation of labour, loss of local cultures, abandonment of rural areas…) and economic problems (unfair distribution of value…). The potential positive contributions of more localised food systems are coming to the fore.

There are clear links with the somewhat controversial concept of multifunctionality (OECD 2000; Van Huylenbroek and Durand 2003). There is recognition of positive and negative externalities related to modern production methods.

Agriculture and food are a special field of analysis for issues related to place. Aspects such as microclimate, soil characteristics, local knowledge, consumers' knowledge are not easily transferable. Today the role of place and local specific resources (genetic resources, knowledge, etc.) is widely recognised in multidisciplinary approaches. This is true not only with regard to technical issues, production processes and management systems, but also with regard to food quality, which allows for product differentiation (e.g., origin products, Geographical Indications). The French concept of "terroir" (Casabianca et al. 2011) summarises the special contribution of place to food quality: natural and human factors are closely connected thanks to a long-term process of mutual adaptation that transforms their nature.

2.2 "Local food" quality and "local quality" food

At the same time, increasing market differentiation and the emergence of the controversial issue of "quality" due to market and regulation failures is leaving more and more room to new quality attributes, some closely related to place.

Modern consumer theories inspired by Lancaster (1966) allow for recognition of multidimensionality of (food) quality: not "quality" of food, but food "qualities" coming from different subsets of product characteristics. Local resources (tangible and intangible) play a great influence over food qualities and consumers' perception, and many product attributes can be related to local resources, such as intrinsic material and immaterial characteristics, extrinsic characteristics (effects that production process exerts on society and environment), and service characteristics.

Local food quality is the result of the combination of many quality characteristics, some of which are place-related. In order to set up some reference points in our discussion, two ideal typical typologies of *local food qualities* can be identified on the basis of three main dimensions: the role of local specific resources, the importance of territory for giving the identity to the product, and the type of target markets (see Table 1).

The first typology can be referred to as "*local food*" quality. Quality is based on the proximity between producer and consumer: physical proximity ("local food") is a food quality attribute per se entailing other extrinsic and intrinsic characteristics. This "local food" quality typology is a result of criticism of "industrial food" and global supply-chains in their different expressions, and shows an interest in fairness and cleanliness of food (processes and products) and towards more authenticity and freshness. Re-spacing effects are not side-effects of consumers' choice; they are the very effects some groups of consumers are interested in. These effects are recognised as closely related to the building of re-connected modes of food provision: it is not only physical proximity that matters, but also organisational, social, political, and cultural proximity between producers and consumers.

The second typology can be referred to as "*local quality*" food. In this case we can speak of a food quality specifically related to a specific place, thanks to the incorporation of natural and human resources of that specific place into the final product. We can speak of "Origin Products" as products originating from a delimited territory where a noted quality, reputation or other characteristic of the good is essentially attributable to its geographical origin and the human or natural factors there. The special role that local resources play in building this kind of "local quality" is emphasised in Geographical Indication (GI) products, where the geographical name of the place of production (or another name linked to that place) identifies the product on the market. GI products are origin products named with a "geographical indication".

Markets of Origin products and GI products are not only local, and often a global niche market perspective is necessary to attract consumers interested in specific characteristics and equipped with information, knowledge, and consumption experience to recognize and assess them. The use of a GI as identifier becomes increasingly necessary as a market grows in order to identify the product coming from that specific place as "different" from any other.

2.3 Local food quality and local resources

There are important differences between the two ideal typical typologies in the role of local resources.

In the "local food" quality typology, local resources play no specific role with regard to product specificity, except for their location. Re-localization of food production-consumption chains is envisaged as an opportunity for mobilising local resources and re-activating local economies, stimulating local development processes and preventing degradation in land use around towns or in local rural cultures.

This re-localization is often blocked where production stages in the supply chain or specific skills are lacking. Erosion of local competences, a side-effect of the spread of a-localised models, is one of many factors hampering re-localization of producer-consumer relationships.

In the "local quality" food typology, specific local resources are necessarily mobilized and embedded in local specific production processes, as otherwise no specific place-linked quality can be revealed in the product. The use of specific local resources allows for a "local quality" of food. Local specific resources, including human skills and know-how, are normally shared by numerous firms and other actors inside the territory, and present some features of common goods.

In both the typologies, food quality comes from the "local" nature of resources employed, although "local" may have different meanings. Local resources, in particular when very specific, can however often be considered as less efficient than the competing standard resources. This is why they are increasingly threatened by extinction. But if we move away from strictly technical terms (technical efficiency), the evaluation of inefficiency may not hold. A broader vision of consumer needs sees them interested not only in price but it also takes account the growing importance of segments sensitive to other product quality attributes (preserving traditions, protecting environment, enhancing social relations) and costs and benefits of these. At the same time, it is thanks to the quality characteristics related to "local", and recognition on the market by big segments of consumers, that local resources can be activated, preserved, and improved.

In both the typologies ("local food" quality and "local quality" food) problems and failures emerge. On resource markets, failure can be caused by a lack of knowledge and skills, lack of organisation and coordination between actors in managing common local resources or in network building. On goods markets, failure can be caused by poor recognition of "qualities" of food in producer-consumer relationships, sometimes as a result of unfair.

These failures are particularly frequent in the case of origin and GI products, the focus of the following sections.

3. Origin products and Geographical Indications: a controversial field

3.1 The Origin product and its "virtuous circle"

The specificity of Origin Products (OPs) comes from their strong link to their territory of origin. There are three dimensions characterizing OPs: the specificity of local resources used in the production process; the history of the product and the production and consumption tradition; and the collective dimension, including presence of a shared knowledge at the local level (Barjolle et al. 1998; Bérard and Marchenay 2004; Casabianca et al. 2011).

The valorisation of OPs is increasingly seen as a powerful tool to enhance, preserve and develop local specific resources, with more expected positive effects on rural development than other kinds of food products. To date, little empirical analysis has been carried out on the positive or negative contributions OP systems may make in these dimensions. Academic studies often refer to an "ideal-type" of OP system, a sort of idealized "virtuous OP system", capable of attaining a fairly high number of beneficial effects (Belletti and Marescotti 2011).

The specificity of local resources determines the peculiarities of product quality attributes, stemming from the "physical" environment where the product is produced, and particularly the paedo-climatic environmental and genetic resources. Nevertheless, natural resources express their potential through growing, breeding, handling and processing practices, managed by local people with their skills. Moreover, there is also a tradition of consuming specific to the place of origin; how and when to eat the product, how to prepare, cook and serve it, and how to evaluate its quality.

By definition, specific local resources are not easily transferable to alternative uses without some loss of productivity. At the same time, they offer opportunity to differentiate the product on the market, and internalize values connected to the place. They allow enhancement, remuneration, reproduction and renewal of local specific resources.

The OP should therefore be considered as a social construct. On the basis of a set of local resources, the behaviour and strategies of individual and collective actors construct the OP over the course of time. The product has to be "validated" by the outside, through the market and/or through public support schemes, which pay for resources and allow local actors to reproduce these resources. In this way, a "virtuous" circle is activated (Belletti and Marescotti 2011); the process is able to close and effectively achieve the reproduction and renewal of the resources used in a "complete" production process.

The ability to create value from an OP via the market mechanism, and the allocation of this value between different firms participating in the supply chain, have a territorial and a collective basis.

As well as specific local resources, the territorial basis is provided by the name of the product which contains a geographical reference. This geographical name is used as the main communication leverage to market the product to consumers. It uses long-term reputation acquired through repeat purchases and the maintenance of the promise of quality (Belletti 2000).

The collective basis is given by local actors' long-term contribution to the definition, evolution and maintenance of the link between the product and its territory. They thus gain the exclusive right to use the geographical name of the product. Value creation by the OP should be conceived of as a local qualification process, or a social construct. By this construct, local actors (producers and other stakeholders) manage the link between product quality and territory, and reach a dynamic agreement on how to link the product to society (consumers and, more in general, citizens) on the basis of certain conventional rules (De Sainte Marie and Casabianca 1995; Tregear et al. 2007).

Thanks to its deep roots in the local context, the functioning of the OP production system implies direct economic effects connected to the marketing of the product, and indirect effects on the stock of territorial capital.[2] Other local and non-local production systems may also benefit from this accumulation of capital, thus conditioning the quality of life of the local community in a way that goes beyond the actors directly involved in the OP production process.

The activation of a virtuous circle may be hampered by obstacles and failures related to three main areas:

a) Technology. The availability of technologies that are alternative to the traditional ones (due also to the success OP gains on the market) may increase the opportunity cost of using local specific resources. These may thus be replaced by more generic ones, loosening the link between the OP and the locality. Iin the same way, if local firms are not sufficiently aware of the role and need to use traditional production technology, they may fail to transmit this information to younger generations. So specific resources have to be replaced by generic ones.

b) Collective action. Because of the collective nature of OPs, the activation and development of the qualification process should be made through collective action. But lack of skills and information, conflicts between actors, the co-presence of different and even divergent actors' interests and aims about the valorisation, may lower levels of coordination among local actors towards a common qualification process. Moreover, power imbalances and a lack of democracy among actors can lead to uneven gains and tensions inside the local production system.
c) Market failures. Failures of the market mechanisms can undermine the remuneration of the OP specific resources. These failures can occur both on the consumer market as a result of limited and asymmetric information, and on the intermediate market as a result of imperfect competition. As a consequence, unfair imitations of the OP can crowd-out the "original" OP and generate unfair value distribution between supply-chain firms. In addition, not all values incorporated in the OP can be successfully communicated to consumers. The higher the existence option values[3] expressed by the OP (e.g., linked to a specific local breed), the more difficult full remuneration of OP values.

All these failures threaten the effective reproduction and renewal of specific local resources.

3.2 Geographical Indication protection and the link between product and territory

Geographical indications, defined as identifiers of the product expressed by geographical names or other words or symbols very specific of their territory of production, are but one of the tools that may allow the qualification of the OP both inside the local production system and on the market. In many legal systems, GIs are recognised and protected against misuse and fraud by means of legal instruments and in some cases by special protection schemes.

The concept of Geographical Indication is defined in the TRIPS Agreement (Art. 22.1) as follows:

> Geographical indications are, for the purposes of this Agreement, indications which identify a good as originating in the territory of a Member, or a region or locality in that territory, where a given quality, reputation or other characteristic of the good is essentially attributable to its geographical origin.

The main question here is: what is the role of local resources in the wide and fuzzy concept of "geographical origin" cited in TRIPS? The concept of "geographical origin" varies depending on different national cultures, as well as on different conceptions of "food quality". Analysis therefore needs to be "contextualised". The European Union, with its big component of Mediterranean countries (France, Italy, Spain) can be considered one of the most "complete" interpretations of the concept of territory around the world.

European Union Regulation 510/2006 provides for a legal protection of Geographical Indications from imitations and abuses on the market. The Regulation introduces a distinction between Protected Designation of Origin (PDO) and Protected Geographical Indication (PGI).[4]

For a PDO there must be an objective and exclusive link between product quality and its geographical origin. For a PGI product, the link with the geographical area does not need to be "essential or exclusive" but has to be causal. In this sense, it is sufficient that the features or the reputation of the product are 'attributable' to the geographic origin.

Actors interested in applying for a PDO or PGI registration have to demonstrate the (more or less strict) link between the quality characteristics of the product and its place of origin. Moreover, and more important, actors present the Product Specification (PS), which sets out the conditions that producers must respect to in order to gain PDO or PGI recognition. The PS has to contain all the technical information regarding the product and is the reference point for assessing that producers using registered names are complying with the rules.[5]

3.3 Geographical Indications and the role of local resources

Notwithstanding the clear and apparently strict rules that require the demonstration of a (more or less) strong link between quality characteristics of the good and its place of origin, and the reference to "natural and human factors", no more is specified about the nature of resources to be cited in order to obtain a PDO or a PGI.

In the case of PGI, it is sufficient to demonstrate a certain reputation of the name of the product in association with its place of production (or, at least, a part of the production). In the case of the PDO, where the link is closer and determines the very quality of the final product, reference is made to natural and human factors without any other qualification ("*essentially or exclusively due to a particular geographical environment with its inherent natural and human factors*"). It is therefore left to local actors to provide evidence and select which (if any) human or natural resource affect product quality.

In theory, according to the terms of the Regulation, an application for PGI status may be made in the absence of local resources, if there exists a reputation inherited from the past.

Moreover, the way the Regulation has been implemented at single Member State level across the EU varies widely in terms of the procedure set by national authorities and documentation required from applicants (Sylvander 2004). There are many differences regarding how to demonstrate the link, supporting evidence and admissible resources. This flexibility has led to various interpretations of the link product-production process-local specific resources, only made more various by the fact that implementation procedures have changed over time in single Member States.

In this context, much of the link between product quality and local resources has come to be defined by local actors themselves. On the basis of local production traditions and knowledge, it is they who set out the conditions of production and the kind, quantity and quality of local specific resources in the Product Specification (PS).

This is a very delicate and complicated issue, as stakeholders participating in the building up of the PS may display diversified if not conflicting interests about how to shape the rules. PS are becoming increasingly "loose" as regards the way and extent local resources are incorporated. Stakeholders usually justify their choices citing market opportunities, lowering production costs, making flexible the production process, use of alternative technologies, improvement of organisation, or even the exclusion of rival firms from PDO/PGI on the basis of some principle of legitimacy.

Depending on the way all these different interests are mediated in the PS, the link between the product and local resources is strengthened or weakened.

4. From "local quality" food to activation and reproduction of local resources: the example of genetic resources

"Local quality" food can be considered as a driving force for orienting local resources mobilized by the supply chain. Actors gathered around the product are impacting on the management of associated specific resources, and this impact needs to be analyzed in terms of resources renewal and ability to close the virtuous circle.

The incorporation of a resource in the production and enhancement of "local quality" food can have several consequences on the future life of the resource itself. "Local quality" food is a strong factor affecting the way activities are located, as well as the characteristics, the quantity, and even the nature of resources. As an example of such issues, we focus on genetic resources and changes when these are included in a "local quality" food dynamic (Verrier and Bouffartigue 1993; Lambert-Derkimba 2007).

4.1 Local breed as a local resource

In a very general meaning, a "breed" refers to a group of domestic animals, originated and maintained by man and having a clearly defined set of characteristics. The classification of breeds is frequently based on the size of the population concerned: "main" breeds have a great number of animals while "minor" breeds present less animals and "endangered" or "rare" breeds have difficulty maintaining genetic stock.

Very few official definitions are available of "local" breed, although the term is often used to identify a less common genotype. Traditional use of localized genetic resources has led to the term "heritage" breed; these are traditional livestock breeds bred before the drastic reduction of breed variety caused by the rise of industrial agriculture (see www.sustainabletable.org).

Heritage animals were bred over time to develop traits that made them particularly well-adapted to local environmental conditions. Breeds used in industrial agriculture are bred to produce big quantities of milk, gain weight quickly, or yield particular types of meat within confined facilities. Heritage breeds are generally better adapted to withstand disease and survive in harsh environmental conditions, and can be better suited to living on pasture.

These livestock breeds also serve as an important genetic resource; when heritage breeds become extinct, their unique genes are lost forever and can no longer be used to breed new traits into existing livestock.

Therefore, by keeping heritage livestock breeds, sustainable farmers not only maintain variety within livestock populations, they also help to preserve valuable traits within the species so that future breeds can endure harsh conditions.

There is no official definition or certification for "heritage" animals, but for a livestock breed to be truly heritage, it must have unique genetic traits and also be raised in a traditional farming and breeding system. Heritage animals are well-suited to traditional breeding farms since they are able to survive without temperature-controlled buildings and the constant doses of antibiotics administered to the commercial breeds raised in a standard breeding system.

"Local breed" (or "regional" breed), may mean a heritage population well described and managed, with individuals mainly located in the same area (in general its native area), although a few individuals may be bred in another areas (Audiot et al. 2005a). Many heritage breeds are also "local" breeds because their performance is insufficient for competing with selected or more specialized breeds (Verrier et al. 2005).

Such local breeds are often included in the product specification (PS) for GI products in order to specify their anchorage in a territory (Roncin 2000). This inclusion may be compulsory for all the breeders wishing to produce according the rules, or compulsory for a certain proportion of the flock in the production unit or at the GI level (Lambert-Derkimba et al. 2006). Empirical case studies may produce useful observations to supply our theoretical framework with relevant questions on how to close the virtuous circle, and on the consequences of closing the circle for renewing local resources.

4.2 Recovering local breeds through local food dynamics: some case studies

"Camembert de Normandie" cheese and Normande cattle breed

Camembert de Normandie is a well-known PDO French cows' milk cheese from Normandy. It has been heavily industrialized in recent decades and big changes have occurred in the genetic structures of the herds. As farmers had no particular requirements, the local cattle breed, Normande, very famous for the high quality fat content of the milk, has become much less common nationally and also in its native area where it has been progressively replaced by the Holstein, more productive and with lower production costs (Vissac 2001).

Recently, the local PDO body, in agreement with French INAO,[6] decided new requirements on the proportion of Normande cattle milk for

PDO Camembert de Normandie. The same occurred for the three other PDO cheeses of Normandy: Livarot, Pont l'Evèque and Neuchâtel. The PS have gradually obliged farmers to re-introduce or to reinforce the numbers of Normande breed in their herd. Increasing proportions over time are defined at PDO level. These measures are obviously ensuring that the local breed does not decrease or disappear. The local breed is confirmed as a pillar of "local quality" food and not only for the picture on the label.

Not all local actors agreed with this vision of the local food. The main tension concerned not so much the use of the local breed, but focused on the issue of raw vs. pasteurized milk for cheese production. Some actors wanted to combine all possible elements into a very coherent orientation of the local food (increase of local breed proportion and exclusive use of raw milk). Others were more interested in the collective reputation of the cheese, where marketing would use the Normande without there being any requirement for it to be present in the herd. They wanted to reduce production costs and health risks by using filtered or pasteurized milk. In this battle, the crucial point was the intermediate market and the requirements for cow milk to be used for these cheeses. Finally, the PDO body received an agreement from INAO on the increasing proportion of the local breed and the obligation to use raw milk. Genetic resources here were considered as part of the strong territorial anchorage of the product, and the decision is giving new importance to the Normande breed in the new PS.

The Gascon pig breed and Bigorre dry cured pork

This breed is one of the last surviving local pig breeds in France. Seemingly doomed to disappearance in the 1970s, it was saved over the following twenty years by a national program with public support. All over France, breeders helped to rear the breed and ensured its survival. In the 1990s, with other animals scattered over France, a nucleus was re-localized in a traditional area in the French Pyrenees, on the basis that Bigorre was the "cradle" of the breed (Sans et al. 2011).

This relocation was part of a local food dynamic trying to create a new supply chain around the use of this particular breed. The strategy was based on the breed as a main differentiation factor and, after several years of constant effort, application was made for a PDO "Noir de Bigorre" for fresh pork meat and cured products. (The pig has a black coat.)

There were several consequences. The breed was relocated, but the processing knowledge for producing dry cured hams was not. So local processors are using basic industrial knowledge to create a completely new product based on the ancient breed. This hybridization needs to be analyzed in terms of requirements made of the animals. Carcasses and back fat content, characteristic of the breed, are not well accepted.

Another important issue concerns the breed. The local dynamic is splitting, with two associations for managing the breed at national level (Audiot et al. 2005b). The pigs located in Bigorre are oriented by the PDO application and managers are selecting a type of pig more in line with the Iberian model. They have introduced new criteria in order to achieve this objective. It can be assumed that the PDO application is modifying the genetic resource and adding new interpretations of tradition. The rest of the breed maintains the previous orientation, and the diversity required by conservation programs, but there is no longer any contact between animals from the two different colonies, officially for health reasons. As a consequence, there are now issues to be resolved for breed governance and management at national level.

Corsican pig breed and Corsican dry cured ham

This breed is another French local pig breed, but its history is different. Still bred in its native area, it is the basis of an extensive livestock system where rustic animals go out to feed in natural forests on acorns and chestnuts. Over the decades, it has been object of several crossbreeding attempts without any collective breed management being tried. The processed pork products are well known in this tourist region. Strong demand for the products supports farmer income in the mountainous parts of the island.

During the late 1990s, the breed was saved by a regional association (Casabianca et al. 2000) and was officially recognized in 2006. On the basis of this success, an application for PDO status for "Corsican" dry pork products was sent to the French INAO. This necessitates the exclusive use of the local breed as an element of territorial anchorage. This requirement is a strong motivation for saving the local breed. Farmers are encouraged to re-enforce its numbers in the herds in order to conform with future PDO requirements. Today numbers are rising, even though the PDO application is not yet completed.

Some effects are already observed as in particular the question of name and property rights: in order to avoid any confusion (Corsican pig could be reared out of the PDO specification for Corsican products), Corsican pig breed had to be renamed as "Nustrale" when "Corsica" has been dedicated to the product on the market for reputation reasons. Such obligation has been observed in several cases and we must notice that tensions can appear among the local actors between the name of the breed and the name of the product to be protected. In this case, the modification has been possible because breed managers are also interested in PDO application and agreed for changing the name.

Another effect concerns animal characteristics and, as a consequence, the criteria for breed management. Some breed managers consider slow growth as a main characteristic of the breed while PDO promoters are more interested in volume and production costs (Lambert-Derkimba et al. 2011). So the criteria of the breed became a crucial question. As in the previous case, the PDO application is shifting the criteria and transforming the breed management.

Finally, thanks to the PDO application, the breeders' market has increased at regional level, with a huge new demand for the breed. The low level of supply is having negative effects on breed management, the number and the price of the certified animals. This intermediate market is an indicator of the closing circle where a local resource is activated and improved by the PDO project.

4.3 Elements for discussion

Genetic resources, such as a local breed, are not permanently localized resources, and can always be re-localized in other places. Spatial dynamics are affecting genetic resources because the management depends heavily on the effective location of the animals. Normande cattle are revitalizing in their area of origin where the breed was losing position. The decision of PDO cheese managers and the French authorities has given a role to breed managers in "local quality" food decision-making. Reallocation of a resource is modifying the local governance and the decision making. As a consequence, the Normande cow, so beautiful on the label with its photogenic coat, is clearly interpreted as a symbol of the territorial anchorage of the cheese.

Within the local food chain, only a fraction of the producers promote a complete vision of the cheese, including in particular raw milk (with associated knowledge for managing the process) and the local cattle breed. They are moving towards closing the circle. The producers who were against this orientation stopped producing the PDO cheese as a consequence of the French INAO decision, but have remained powerful in the cream and butter PDO, which uses mainly pasteurized milk from Holstein cows. So in the same area two kinds of PS for PDO products co-exist. The PDO cheese actors are trying to close the circle and the local breed is one of the more visible elements of circle closure. The PDO butter and cream actors are benefiting from the strong reputation of the production place but they are not activating or renewing any local resource. In this case, closing the circle creates conflicting trajectories and increasing tensions.

The relocalization of one colony of the Gascon pig breed is giving new opportunities for the breed in the area. But at the same time, this movement is bringing new risks for breed governance. On one hand, the products are benefiting from a collective identification with the geographical name Bigorre. On the other hand, outside the area, products from the breed are still sold directly using the name of the breed. The same resource is under tension as the breeders no longer constitute a single community. The case of the Noir de Bigorre PDO application can be seen as a reinvention of tradition (Hobsbawm 2006), the breed being re-localized without local practices. But the knowledge originally associated with the genetic resource has been lost. In the new location, cognitive resources on breeding practices and processing procedures are lacking. This set of resources, material, symbolic, and cultural is a system. So spatial dynamics cannot always ensure the coherence of such a system, and knowledge must be considered in full as a local resource.

The example of Corsica provides another useful lesson: the way of enhancing the resource is modified by the local food dynamics. This enhancement has led to tensions between managers and users of a resource. Who is legitimised to plan the future of a local resource? In this case, local food actors have power over the resource thanks to their connection to the market and the strong social capital they provide. The food project is becoming more important than the breed project.

This transformation is also modifying time references. The breed and inter-generation transmission is a long-term project while the food is more influenced by a short-term dynamics and shifting market sensitivity. The PDO is activating the local resource and may also affect its long-term orientation.

So the use and enhancement of local resources in the Origin Product process of GI protection can have many controversial effects, with different outcomes affecting the closure of the virtuous circle. The circle can usually be closed by regulating the intermediate markets of raw material produced by livestock activities. The characteristics of the milk or the meat to be processed are a key point. Using a local breed often necessitates new rules to give new added value to the raw material, in particular because local breeds may be unable to compete with with more productive breeds (Hauwuy et al. 2006).

And closing the circle is transforming local resources. The back effect on the resource itself is quite clear in most of the cases (Lambert-Derkimba et al. 2010). Including the local breed in a PDO system has huge consequences on the breed, on the criteria for its management and on the functioning of the breeders market.

5. Conclusions

The critical issues identified by comparing the theoretical model with the empirical evidence on genetic resources can be seen generally in cases of activation of tangible, intangible, symbolic and cultural local resources in the valorization processes of Origin Products, an expression of "local quality" food.

A central issue relates to the connection between resource activation and their transformation into assets. In this process, downstream actors in direct contact with the final food market are benefiting from a higher stock of social capital (Shorthall 2008) than raw material producers. Upstream, farmers are more directly linked to the management of local specific resources. These resources are subject to appropriation by several types of actors, raising issues of ownership and governance. It can happen that resources are governed by the logic of the market logic, and become unable to maintain their own nature and orientation.

This process of resource activation also leads to a process of selection. Because the relationships between the local food project and the resource managers are unequal, there are risks of exclusion. These risks are more evident in the case of protected GI products because of the need to formalize rules in the Product Specification. If the use of one rare resource is mandatory in the Product Specification, this becomes a way to eliminate actors who are less endowed or face problems in compliance with the rules. In a more general view, the valorisation of Origin Products may have perverse effects when the required resources are not fully available for all potential actors. In such cases, activating and renewing local resources through the origin food system does not systematically lead to moves towards closing the circle and making it "virtuous".

A set of priorities for further research needs to be defined in the field linking local food dynamics and local resources.

Cognitive and relational dimensions have to be addressed more clearly when local food quality is studied as a way of activating local resources. More attention should be paid to how local actors build up heritage; there needs to be a focus on inter-generational links, particularly regarding biodiversity and local knowledge resources.

Considering the lack of theoretical advances in this field, a special focus could be put on the legitimizing process of the managers of common goods as well as the "drivers" of a local food system. Capturing a name for exclusive, even if collective, use is a big responsibility and the way the owners are validated for such exclusivity should be studied in more depth. More attention is needed on the empowerment of local actors through collective action and institutional support in the local food and origin food dynamics.

As a final issue, the distinction we introduced between "local food" quality and "local quality" food should be analyzed in a wider view, considering if these two categories are in competition on the market, and if and to what extent complementarities could be developed.

In conclusion, the theoretical model of the Origin Product virtuous circle shows that for "local quality" food, activation and renewing of resources are key factors in enhancing public interest and justifying the intervention and support of public authorities. The way specific local resources are incorporated (and codified) in the Origin Product process could be the major element for closing the circle and ensuring that it is virtuous for the sustainability of the "local quality" food system.

Two main issues are emphasized in the chapter.

The first deals with the role of public authorities and policy-makers in balancing the interests of each type of actor. In particular, in the case of registered GIs, the Product Specification needs to be drawn up under transparent decision-making on the trade-offs. It needs to be acceptable by all parties and should not discriminate against actors in an unfair mechanism. This means giving room to opposing voices. The need for democracy is at the core of the Origin Product system.

The second issue is the importance of collective action for ensuring cohesion at local level and managing the Origin Product, the registered GI (if any) and the specific resources as a (local) common good. Different, even divergent, interests can of course exist, so building up consensus among local actors and generating shared strategies is a real necessity. Here too, public authorities can support local actor coordination for example in defining rules for GI recognition in the interests of the public good.

Notes

1. See for example Goodman (2004), Brunori et al. (2010), Watts et al. (2005).
2. Territorial capital can be defined as the set of assets of different nature which characterize territories. Assets are given not only by natural resources and private / public fixed capital stock, and human and cultural capital, but also by social, institutional and symbolic capital (Camagni 2008).
3. Existence and option values are part of non-use values, as they do not require utility to be derived from direct use of the resource. In the first case the utility

comes from simply knowing that the resource exists; in the second one the utility is given by the preservation of the option to benefit from the resource in the future (Pearce and Turner, 1990).

4. The definitions of PDO and PGI are given in Article 2 of Regulation 510/2006. They are as follows:

'Designation of origin' [PDO] means the name of a region, a specific place or, in exceptional cases, a country, used to describe an agricultural product or a foodstuff a) Originating in that region, specific place or country; b) The quality or characteristics of which are essentially or exclusively due to a particular geographical environment with its inherent natural and human factors; and c) The production, processing and preparation of which take place in the defined geographical area.

'Geographical indication' [PGI] means the name of a region, a specific place or, in exceptional cases, a country, used to describe an agricultural product or a foodstuff: a) Originating in that region, specific place or country; b) Which possesses a specific quality, reputation or other characteristics attributable to that geographical origin; and c) The production and/or processing and/or preparation of which take place in the defined geographical area.

5. The Product Specification is the determining factor in obtaining registration. As such, the Product Specification must comply with a set of rules stipulated in Article 4 of Regulation No 510/2006 and must include: The product name; A description of the product, including raw materials and "physical, chemical, microbiological or taste characteristics"; The definition of the geographical area; Evidence that the product in the market bearing the PDO/PGI originates in the defined geographic area; A description of how the product is obtained including relevant information on production, processing and packaging methods; Details of the link between the geographical area and the product's quality, reputation or characteristics; Names and addresses of bodies verifying compliance with the specification; Any specific labelling requirements; and, Any requirements laid down by Community or national provisions.

6. INAO is the French public institution in charge of the first step of recognition process for GI applications.

References

Audiot, A. et al. (2005a). Populations animales locales et produits de qualité: comment valoriser transforme la ressource?. *Les Actes du BRG* 5: 577-592.

—. et al. (2005b). La valorisation des races porcines locales du sud de la France continentale. *Ethnozootechnie* 76: 97-109.

Barham, E. and Sylvander, B. (eds.) (2011). *Labels of Origin for Food. Local development, Global Recognition* Wallingford, UK: CABI International.

Barjolle, D et al. (1998). *Le lien au terroir. Bilan des travaux de recherche*. Lausanne: ETH, Isitut d'économie rurale.

Becattini, G. (1989). Riflessioni sul distretto culturale marshalliano come concetto socio-economico. *Stato e mercato* 25: 111-128.

Belletti, G. (2000). Origin labelled products, reputation, and etherogeneity of firms. In Sylvander, B. et al. (eds.), *The socio-economics of origin labelled products in agro-food supply chains: spatial, institutional and co-ordination aspects*, vol. 1, 239-260. *Actes et Communications* 17, Paris: INRA.

Belletti, G. and Marescotti, A. (2011). Origin products, GI special protection schemes and rural development. In Barham, E. and Sylvander, B. (eds.), *Labels of Origin for Food. Local Development, Global Recognition*, 75-91. Cambridge, USA: CABI International.

Benko, G. and Pecqueur, B. (2001). Les ressources des territoires et les territoires des resources. *Finisterra* XXXVI: 717-19.

Bérard, L. et Marchenay, P. (2004). *Les produits de terroir. Entre cultures et règlements*. Paris: CNRS editions.

Bérard, L. and Marchenay, P. (2006). Local products and geographical indications: taking account of local knowledge and biodiversity. *International Social Science Journal. Cultural Diversity and Biodiversity* 187: 109-116.

Brunori, G. (2007). Local food and alternative food networks: a communication perspective. In Amilien, V. and Holt, G. (eds.), From local food to localised food, *Anthropology of Food* S2, March 2007.

Brunori, G. et al. (2010). Co-producing transition: innovation processes in farms adhering to Solidarity-based Purchase Groups (GAS) in Tuscany, Italy. *International Journal of Sociology of Agriculture & Food* 18 (1): 28-53.

Camagni, R. (2008). Regional Competitiveness: Towards a Concept of Territorial Capital. In Capello, R. et al., *Modelling Regional Scenarios for the Enlarged Europe*. Berlin, Heidelberg: Springer.

Capello, R. (2009). *Recent Theoretical Directions and Future Challenges in Regional Economics*. Paper presented at 3rd Central European Conference in Regional Science—CERS, 2009.

Casabianca, F. et al. (2000). L'amorce d'une gestion collective de la race porcine Corse. Construire un standard et élaborer les contrôles des reproducteurs. *Options méditerranéennes: Tradition and innovation in Mediterranean pig production* 41: 23-34.

—. et al. (2011). Terroir et typicité. Un enjeu de terminologie pour les Indications géographiques. In Delfosse, C. (ed.), *La mode du terroir et les produits alimentaires*, 101-117. Paris: Les Indes Savantes.

Cerdan, C. et Fournier, St. (2007). Le système agroalimentaire localisé comme produit de l'activation des ressources territoriales. Enjeux et contraintes du développement local des productions agroalimentaires artisanales. In Gumuchian, H. and Pecqueur, B. (eds.), *La ressource territoriale*, 104-123. Paris: Economica.

Coley, D. et al. (2009). Local food, food miles and carbon emissions: A comparison of farm shop and mass distribution approaches. *Food Policy* 34 (2): 150-155.

Committee of the Regions (2011). Outlook opinion of the Committee of the Regions on Local Food Systems. 88th Plenary session, January 27-28 2011, Rapporteur Ms L.Dwarshuis-Van de Beeke. NAT-V-005. Brussels.

Courlet, C. (2007). Du développement économique situé. In Gumuchian, H. and Pecqueur, B. (eds.), *La ressource territoriale*, 32-44. Paris: Ed. Economica.

de Sainte Marie, C. and Casabianca, F. (1995). Innover dans des productions patrimoniales. Génération d'objets techniques et émergence de règles dans les processus d'organisation. *Cahier d'Economie et Sociologie Rurale* 37: 150-175.

Dunne, J.B. et al. (2010). What does "local" mean in the grocery store? Multiplicity in food retailers' perspectives on sourcing and marketing local foods. *Renewable Agriculture and Food Systems* 26 (1): 46-59.

Duram, L. and Oberholtzer, L. (2010). A Geographical Approach to Place and Natural Resource Use in Local Food Systems. *Renewable Agriculture and Food Systems* 25: 99-108.

Fonte, M. and Papadopoulos, A.G. (eds.) (2010). Naming Food after Places. Food relocalisation and knowledge dynamics in rural development. Aldershot: Ashgate.

Goodman, D. (2004). Rural Europe Redux? Reflections on Alternative Agro-Food Networks and Paradigm Change. *Sociologia Ruralis* 44 (1): 3-16.

Hand, M.S. and Martinez, S. (2010). Just what does local mean?. *Choices* 25 (1): 13-18.

Harris, E.M. (2010). Eat Local? Constructions of Place in Alternative Food Politics. *Geography Compass* 4: 355-369.

Hauwuy, A. et al. (2006). Conséquences de la présence de filières fromagères bénéficiant d'une Indication Géographique sur l'activité agricole des zones considérées: l'exemple des Alpes du Nord. *INRA Productions Animales* 19: 371-380.

Hodgson, G.M. (1998). The Approach of Institutional Economics. *Journal of Economic Literature* 36: 166-192.

Hobsbawm, E. (2006). *L'invention de la tradition.* Paris: Éditions Amsterdam.

Kneafsey, M. (2010). The region in food—important or irrelevant?. *Cambridge Journal of Regions, Economy and Society* 3: 177-190.

Lamara, H. (2009). Les deux piliers de la construction territoriale: coordination des acteurs et ressources territoriales, Développement durable et territoires. http://developpementdurable.revues.org/index8208.html.

Lambert-Derkimba, A. (2007). *Inscription des races locales dans les conditions de production des produits animaux sous AOC: enjeux et conséquences pour la gestion collective des races mobilisées.* PhD, AgroParisTech.

Lambert-Derkimba, A. et al. (2006). L'inscription du type génétique dans les règlements techniques des produits animaux sous AOC: conséquences pour les races animales. *INRA Productions Animales* 19: 357-370.

—. et al. (2010). Consequences of the inscription of local breeds in protected designation of origin cow cheese specifications for the genetic management of the herds. *Animal* 4 (12): 1976-1986.

—. et al. (2011). Les tensions entre ressources génétiques locales et ancrage territorial des produits. Le cas du porc Nustrale et du projet d'AOP charcuterie de Corse. *Economie Rurale* 322: 39-49.

Lancaster, K.J. (1966). A New Approach to Consumer Theory. *Journal of Political Economy* 74: 132-157.

Larson, J. (2007). *Relevance of geographical indications and designations of origin for the sustainable use of genetic resources.* Roma: Global Facilitation Unit for Underutilized Species.

Linck, T. (2007). Le dilemme de la gestion patrimoniale. L'exclusion, condition et obstacle à l'appropriation collective. *Économie Appliquée* LX (3): 177-198.

Mathijs, E. et al. (2006). *Instruments and institutions to develop local food systems, Scientific Support Plan for a Sustainable Development Policy.* Part. 1. Sustainable production and consumption patterns. Brussels: Belgian Science Policy.

OECD (2000). *Multifunctionality: towards an analytical framework.* Directorate for Food, Agriculture and Fisheries. Trade Directorate COM/AGR/APM/TD/WP(2000)3/FINAL 2000 December.

—. (2008). *Environmental Performance of Agriculture in OECD Countries Since 1990.* Paris: OECD.

Pearce, D.W. and Turner, R.K. (1990). *Economics of natural resources and the environment.* Exeter: BPCC Wheatsons Ltd.

Pecqueur, B. (2006). Le tournant territorial de l'économie globale. *Espaces et sociétés* 2: 17-32.

Pretty, J.N. et al. (2005). Farm costs and food miles: An assessment of the full cost of the UK weekly food basket. *Food Policy* 30 (1): 1-19.

Prost, J. et al. (1994), La certification des produits, un levier pour le développement de l'élevage. La dynamique de l'AO «Brocciu Corse». *Etudes et Recherches sur les Systèmes Agraires et le Développement* 28: 143-156.

Requier-Desjardins, D. (2009). Territoires – Identités – patrimoine: une approche économique?. *Développement durable et territoires* (Dossier 12: Identités, patrimoines collectifs et développement soutenable) http://developpementdurable.revues.org/7852.

Roncin, F. (2000). Les enjeux de la conservation des races locales, position du problème pour les signes de qualité: cas des appellations d'origine. In *Races locales et biodiversité animale; Enjeux et appétits*, 147-157. Actes du colloque FAIR, Rambouillet, France.

Sans, P. et al. (2011). L'origine des viandes et des produits carnés: le terroir reconstruit? In Defosse, Cl. (ed.), *La mode du terroir et les produits alimentaires*, 235-259. Paris: Les Indes Savantes.

Shortall, S. (2008). Are rural development programmes socially inclusive? Social inclusion, civic engagement, participation, and social capital: Exploring the differences. *Journal of Rural Studies* 24: 450-457.

Sylvander, B. (2004). *Dolphins: Final Report. Synthesis and Recommendations*, Le Mans: Dolphins UE FP5 research project.

Thévenod-Mottet, E. (2010). Geographical Indications and biodiversity. In Lockie, S. and Carpenter, D. (eds.), *Agriculture, Biodiversity and Markets*, 201-212. London: Earthscan.

Van Huylenbroek, G. and Durand, G. (eds.) (2003). *Multifunctionality: A new paradigm for European agriculture and rural development?*. Aldershot: Ashgate.

Verrier, E. et al. (2005). Conservation and values of local livestock breeds: usefulness of niche products and/or adaptation to specific environments. *Animal Genetic Resources Information* 36: 21-31.

Verrier, E. and Bouffartigue, B. (1993). Les AOC, élément du maintien des races régionales? L'exemple des Alpes du Nord et réflexions sur d'autres situations. *Ethnozootechnie* 52, 33-42.

Vissac, B. (2001). *Les vaches de la république*. Paris: INRA Edition.

Watts, D.C.H. et al. (2005). Making reconnections in agro-food geography: alternative systems of food provision. *Progress in Human Geography* 29 (1): 22-40:

CHAPTER FIVE

GLOBALIZATION/FRAGMENTATION PROCESS: GOVERNANCE AND PUBLIC POLICIES FOR LOCALIZED AGRI-FOOD SYSTEMS

GERARDO TORRES SALCIDO
AND JOSÉ MUCHNIK

Abstract

Globalization processes and new forms of political, economic and social fragmentation are developing at the same time. This chapter tackles a double paradox. The first is the paradox of consumption fragmentation: in a world characterized by the standardization of goods and services, the emerging consumer is demanding products with specific characteristics associated with a territory, which indicate that it belongs to a place and a culture. The second is the paradox of collective actions vs. social exclusions: the growing demands on producers to ensure authenticity of a product and compliance with environmental regulations through certification mechanisms, most often foreign to the producer, cause new inequalities and asymmetries in the organization and collective action both inside and outside communities. This chapter is divided into three parts that analyze these paradoxes, in relation to governance; their conceptual relationships; and their importance for public policy.

This work is supported by studies and empirical research carried out in Latin America, particularly Mexico (Project: "Sistemas agroalimentarios locales en México. Identidad territorial, construcción de capital social e instituciones" [Local Agri-food Systems in Mexico. Territorial Identity, Building of Social Capital and Institutions])[PAPIIT IN301708-UNAM; 80333-CONACYT] and Argentina (Proyecto Interra del programa SYSTERRA [Interra Project, SYSTERRA Program] [ANR-09-STRA-04], Agence Nationale de la Recherche, France).

1. Introduction

Recent literature on local agri-food systems (SYAL, for the acronym in French, from "Systèmes agro-alimentaires localises") SYALhas been emphatic in showing the virtues and contributions of these systems in understanding the multifunctionality of the rural setting (Rodríguez-Borray and Requier-Desjardins 2006), the specificity of these systems and their territorial link (Bengoa 2007; Requier-Desjardins 2001), as well as the processes of product certification (Oyarzún 2005, 24). Attention has also been called to the possibilities represented by this approach for implementing public policies to fight rural poverty (Requier-Desjardins 2006). Emphasis has also been placed on its relation to collective action, associationism and social capital (Torres Salcido and Sanz Cañada 2007) and the use of common-pool resources, which will be of great interest in this work (Requier-Desjardins 2004). However, in the literature there are no detailed studies on the conjunction of the two paradoxical phenomena that involve consumers, sellers, service organizations, food producers and their effect on SYAL: globalization, on the one hand, which is imposing a vertical coordination of production and consumption and, on the other hand, the need to consume products differentiated by seals of authenticity, that is, with territorial characteristics, which is leading to the fragmentation of the manner of consumption and putting pressure on producers through the need to certify products.

The analysis of fragmentation of consumption needs to be framed in a critical analysis of "globalization". "Globalization" in fact is an inexact term as it hides the conflicting process of fragmentation it involves. The political changes of the past 20 years—the emergence/claims of "new territorialities", the acceleration of the economic and financial crisis that started in 2008, and the recent revolts by the "Occupy" movements throughout the world—show just how deep these fragmentations run at different levels. We can also see this double trend in the production of goods and services: on the one hand, we have standardized activities, technologies and products (we can buy the same computer, the same can of beer or the same deodorant in different parts of the planet), but, on the other hand, people are still linked to specific societies, territories, cultures, values and goods, which give meaning to their lives and daily acts.

The paradox of the fragmentation of consumption arises in "affluent societies", dissatisfied with mass consumption, but at the same time besieged by poverty and deep inequalities. The second paradox has to do with SYAL governance, and has its origin in demand itself. The coordination of actors and collective actions, which aim to value territorial resources through responsible or fair consumption, demands processes to certify and defend the products identified with territories. Paradoxically however it may result in exclusion processes, because certification implies (as will be seen later) collective performance by the groups that are most active and have the greatest capacity for transmitting and accumulating capabilities. This paradox is explained by the inability of standardized food, characteristic of mass consumption, to nourish and provide cultural identity to individuals. The processes of standardization pose the question of indifferentiation of individuals, although this is not opposed to their individuation but in fact makes it even more necessary. "The individuation of a singular existence, far from being an isolated fact, is a *linea generationis substantiae*" (Agamben 1990). This explains why globalization and fragmentation happen simultaneously: on the one hand, they are opposites, but on the other they give each other reciprocal meaning. To drink Coca Cola with a *mole poblano* or a *bife de chorizo*, and finish the meal with a "globalized" frozen dessert is not a contradiction in terms, it is a paradox that, on further analysis, allows us to understand the meaning of the evolution we are witnessing.

Man always defines himself with regard to "other one", be it to find similarities or differences. An individual's identity is made up of the set of references that allow that person to be recognized as such, and not be mistaken with another person. These identity references are the result of a social construction and symbolize two levels of relationships: social references (language, place of birth, religion), which define an individual's belonging to a community, and individual references (first name, last name, trade), which differentiate one individual from another; that is to say, identitarian (self-defining) references are the sensitive and memorial bases upon which we build our different ways of inhabiting the world according to different modalities of relating to ourselves and others (Ortigues 1989). In this sense, food is an essential self-defining reference; from the family kitchen to regional cuisines, food is part of identity building for individuals and societies. "Tell me what you eat and I will tell you who you are", said Brillat Savarin (1825) almost two centuries ago, revisiting the old German saying "*Man iss, was man isst*". We can see why the more standardized products there are, the more the consumer needs products with specific territorial characteristics, since food plays a particular role in the individuation/indifferentiation dialectic relationship. To be more explicit, let us say that the more hamburgers and pop corn there is, the greater the need for tortillas and *tlayudas*, *arepas santanderianas*, tamales, *queijo de coalho*, *sopas de chuño*, and bifes de chorizo, to name but a few examples from the vast Latin American culinary diversity.

Standardized products have lost their deep territorial roots, which is why, in the terms of anthropologist Marc Augé, they are like "non-places", spaces in anonymity. A non-place must not be confused with utopia, an imaginary setting laden with hope, but with its opposite, antitopia or dystopia. Augé (2005) gives the name "non-place" to urban spaces, such as shopping malls or airports, conducive to impersonal and anonymous interactions. In literature, mass consumption evokes the non-life of Peter Schlemihl, the character created by German Romantic Adelbert von Chamisso, who exchanges his shadow for an enigmatic bag of wishes. Schlemihl thus loses his traits, the personality which makes him different from his peers; he loses what allows him to interact socially (von Chamisso 1994). The character loses his story through modernity's Faust-like aspect. This literary example explains the paradox from another angle: the paradox of consumption fragmentation in modern society, which ceaselessly pushes towards massification and, as a consequence, to the deterritorialization and loss of contact between local products and the consumer. It gives rise at the same time to the opposite and complementary phenomenon, reflected in the search for a differentiated consumption, which the consumer associates with individuation, his/her roots and history. The consumer thus buys products not only for their physical attributes, linked to territory, but also for their symbolic content, rural lifestyles, landscapes and the sentimental value they transmit.

The second paradox, marked by the opposition and complementarity between collective actions and social exclusion, is related to the dynamics we have just described. The consumer and regulating bodies demand that the producer keep a strict control of the attributes of the food (organoleptic, sanitary, physical traits) through certification granted by regulations, rules and conventions that have been systematized from the observation of traditional production processes and later imposed by exogenous bodies of certification and quality normalization, e.g., the International Organization for Standardization (ISO).

These product qualification processes can become a heavy burden for the producer who is concerned about maintaining organoleptic requirements and symbolic traits, and even heavier if that producer's activity is based on territories that are very different from those in which the consumer lives. Safety issues in recent years have given rise to new "food scares" such as mad cow disease (bovine spongiform encephalopathy) and bird flu (which originated in the H5N1 virus), thus increasing the demands placed upon producers.

In recent decades, advances in sociology and social sciences have brought to light certain elements that help to interpret the crisis of the vertical and mass manners of production and consumption. For Pierre Bourdieu (2000), distinction is one of the incentives for the accumulation of capital, not only in its physical or palpable form, but also as relational knowledge and symbolic capital. The need for this distinction may be manifested in the way in which medium- and high-class consumers qualify products. When Bourdieu states that what distinguishes the bourgeois from other classes is, precisely, distinction, he means the capacity of one part of the population to separate itself from the masses by means of its power of consumption and accumulation of things and activities (tourism, for example) desired by the majority but unattainable for the time being. Notwithstanding, mass consumption is characterized by its expansive power, to the point of socializing the enjoyment of what used to be considered distinguished. The paradox lies in the fact that, on reaching this stage, consumption loses its distinctive characteristics and becomes something generic, and thus there is a need for a constant search for new fields of distinguished or exclusive consumption. One of the risks of distinguished consumption is that it can lead to processes of private appropriation of common goods, such as landscape, natural or cultural heritage and signs of territorial authenticity. In a sense, as pointed out by Ulrich Beck (2006), that risk can lead to new types of inequality that reinforce and strengthen the old and known inequitable structures of modern society. New social proposals for consumption as a part of new lifestyles and consumer identification with the producer's interests may in fact lead to a stratification of subjects and to the individual appropriation of territorial goods.

The paradox stems from the fact that the transaction costs imposed by regulation and certification of production and post-harvest processes, and safety standards required by the consumer and regulating bodies, may lead to non-inclusive organizational processes. These may lead to the creation of new rural elites based on an increase in social capital, with specific abilities and capabilities that cannot be generalized among rural populations. In Latin America, there is strong empirical evidence of the paradox triggered by this new governance of the agri-food system. The production of typical cheeses, fair-trade coffee networks, and the trade and industrialization of products with a strong territorial anchoring (as is the case of the nopal, a cactus vegetable in Mexico) are all taking place in the double paradox of urban consumption. On the one hand, there is an aspiration to distinction through the revival of products with territorial identity and, on the other hand, there are exclusion processes against organizing and building alternatives from the bottom up.

The following section deals with the methodology of defining the role played by governance in the network of concepts surrounding SYAL, and discusses this. It ends by looking at the importance of SYAL for public policies and a new model of development.

2. The method. Clarifications of concepts

This chapter deals with the subject of SYAL under a double focus: (i) as a new way of organization and collective action; (ii) as an analytic category, a concept under construction that allows us to describe these new ways of production organization and consumption. The heuristic method was considered appropriate for the goals of this study. The characteristics of SYAL will be defined by analyzing concepts, using one of those concepts as a common theme. The key function of the concept is to arrange and give a hierarchical structure to the epistemic reflections surrounding the analytic approach. It is important for the construction of a research object to bear in mind the "historic conditions so that we can say 'something' aboutan object, and that different people can say different things so that [said object] will be inscribed in a relational field with other objects" (Foucault 1969). In this sense, in the first place, we start from the assumption that the concept of SYAL appeared in the context of crisis for the dominating agricultural model, and that, as a category, it has spread widely in the last decade among scientists and agents of development interested in rural territories. In second place, governance is established as a nodal concept, an articulator of the network of concepts that are at the source of this new analytic category. And in third place, the characteristics of collective action are analyzed as a construct of governance with a public policy agenda.

As has been mentioned in other publications and forums, the epistemic status of SYAL as a category is given by the evolution of a series of trends and concepts related in their origins by the political economy of space and the national economy which, recovered and elaborated by the Systems Theory, industrial sociology and human geography, create a new way of looking at the articulation of local economies and the role of agri-food cultures (that is, production, exchange and consumption). We will not dwell on the historical background of the SYAL category in this study. For a closer look at the subject, see the literature available (Muchnik 1996; Boucher et al. 2000; Requier-Desjardins 2001; Cascante 2002; Filippi and Torre 2002; Pecqueur 2004; Álvarez Macías et al. 2006; Boucher 2006;

Poméon 2006; Muchnik et al. 2008; Torres Salcido 2009). What is important about this category is the analytic focus, whose utility is being widely debated for challenges such as the rural sphere, city feeding and diet, local development and the generation of public policy proposals imbued with a strong territorial focus.

The rebuilding of the SYAL category and underpinning concepts is useful. They are relevant to the creation of a public policy agenda with its own specificity, different from that of other categories and focuses. These include industrial districts, clusters and local production systems, which are more concerned with the economic dynamics of sprawl. Due to the semantic similarity of the terms, they have mistakenly been confused with SYAL. SYAL studies consider intervention, grounding or grafting of institutions and social relationships important in the operation of local economic activities. So they revive the idea of a "substantive" economy, that is, an economy rooted in social life (Polanyi 2006).

The triple specificity mentioned above—the conjunction of human action, institutions and food culture—is the first difference from those categories and focuses. This triple specificity takes us back to the complexity of territory governance. This concept is networked with other concepts and can aid in understanding the focus we are discussing. The new SYAL governance rises from the crisis of legitimacy faced by the dominant agri-food model and the verticality of public management. The concept itself is controversial. Multiple definitions have arisen with the knowledge that private and public management are complex. Aguilar (2006) has described the historical evolution of the concept of governance and its different variables. It is not the goal of this chapter to review all definitions. In terms of SYAL structure, governance is the process of building agreements to increase wellbeing by managing a territory's tangible and intangible resources. It involves managing, directing and coordinating socioeconomic processes in a specific environmental context, with local institutions and social actors—at micro-level (territory), meso-level (region) and macro-level (global economies or world economies)—who articulate their collective action in terms of the value appropriation of territorial resources or the expectation of wellbeing generated by the valuing of those resources.

Governance is determinant for the insertion of local economies in world economies under collective and largely cooperative schemes. It is not necessarily implied that regions are divided into winners and losers, as they have usually been by approaches on innovative environments or regions of knowledge and that have been the basis for designing clustering or sprawling policies for local production systems.

The conjunction of specificities is the key to thinking of governance in different terms than those set forth by prisoner-type games in which the need for a win-win situation prevails and depends on exogenous factors. In the case of the prisoner's dilemma, the maximization of profit—to save oneself by betraying somebody else, or avoiding punishment by reaching a no-betrayal agreement—depends on the pressure exerted by the policeman-interrogator. In cooperative games, on the other hand, profit depends on collaboration agreements. This does not imply a pragmatic agreement dictated by external factors, but a game of ethics in which the value of acknowledgment constitutes the essence of proximity. In the case of SYAL, this translates into cooperative territorial processes. Governance is the articulation of those efforts to build alternatives to rural poverty and mechanisms that face the deterioration of eco-social systems, through decentralized management. Governance refers to agreements and innovation in three dimensions: a) institutional, building a base in accordance with development; b) social, raising the degrees of trust among actors and achieving an environment conducive to spreading know-how, that is, transmitting knowledge and the territorial identity traits of products; and c) technical, achieving a social construction of technology and the processes of innovation that strengthen relevant findings for producers and consumers in given spatial scales.

So in what sense can governance be a concept that explains the sources and agenda for the SYAL research program? Governance is a political and institutional element that coheres local production systems into a complex system relevant for the constitution and operation of wider systems. In this context, SYAL not only obey self-regulating and self-managing organizations belonging to the administration of common-pool resources (CPR), referred to by Elinor Ostrom (2000), but they involve the interaction of those organizations with the State (legal regulations, economy, central or federal and local governments, education and coercion in a territory). "Complexity" refers to the structuring of relations between society and the environment in a situation of uncertainty, so that it is essential to have processes for self-organization, self-regulation and self-management that lower the costs of externalities. There is also the need for governments and institutions (regulatory, academic, etc.) to be transparent and follow a hands-on approach, in order to guarantee the quality of public management, with regular accountability in order to diminish corruption as much as possible. An ideal type of SYAL would be an agri-food system (production/transformation/services) in a specific territory in which actors try to set up coordination and collaboration processes in partnership terms, with internal management and regulation, but with strong ties to public management and companies. Organizations and associations would have transparent mechanisms for spreading knowledge and distributing benefits and would periodically be held accountable. Since SYAL base their collective action on appropriation and building of tangible and intangible territorial heritage, it is necessary that their action be sustainable, promoting modes of production and consumption that are less harmful to natural and cultural diversity.

This is not about setting up a prescriptive view of SYAL, but a type of governance for these systems. We could use the European context as an example, with olive oil co-operatives in Andalusia, Protected Designations of Origin for cheese and wine in France, and wine co-operatives in Italy, to name but a few. But the abstraction of an ideal type of SYAL is a methodological task. From an epistemic viewpoint, there are no pure SYAL in empiric reality. In Weberian terms, however, systems can have a typical social action as reference. From the elaboration of an ideal type, we can see that there are also self-managing systems and systems introduced by public policies and programs. On the other hand, we find numerous examples of these complex systems that on a daily basis aim at increasing territorial coordination and reducing negative environmental externalities stimulated by a demand model.

Based on the definition of SYAL governance we can specify the degree of coordination between institutions, production characteristics and innovation processes. This methodological procedure is very useful for observing the operation of these systems. In the case of Mexico, the weakness of public management has been widely mentioned as a specific trait that hinders rural development, in contrast to European societies. It has also been pointed out, however, that one of the characteristics of Mexican rural societies is their organizational strength and the persistence of local institutions (if somewhat unfocused), which has provided a greater resilience and openness to innovation. The production and commercialization of organic coffee by indigenous populations of Oaxaca and Chiapas or among small businesspeople in Veracruz, who try simultaneously to certify quality and promote agro-industrial processes in the commercialization of coffee; the production of handmade cheese and traditional liquors or corn tortillas in Tlaxcala by exploiting wastewater from the nixtamalization process, or by making nixtamal with solar energy, are all examples of institutional, technological and social innovation which, together with tourism and gastronomic services, strengthen the multifunctionality of an environmentally sustainable agriculture. We can see similar examples in most regions of Latin America and the Caribbean.

Serious obstacles to integration and innovation can be found in these systems, however. These include public management's lack of interest in development processes and the distance felt by producers from academic institutions, not to mention weak leadership or the vertical structure of product valuation. These obstacles are aggravated by the ethnic and linguistic attributes for which producers are discriminated against in multiple ways by the urban and mestizo society. Another barrier is migration, with the ensuing loss of know-how, and the change in property structure, although sometimes the loss caused by the exodus in search of better opportunities is compensated by sheer will, hard work and the rise of women as a force in the rural environment.

Going back to the problem of system sustainability, it must be assumed that the concept of governance is hierarchically essential in the heuristic method we have chosen, because it is also related to the problems of collective action and public policy. Governance of the complex environment-society takes us to another problem of social sciences and of the operation of SYAL that are related to institutional performance: What to do with a territory's heritage and resources? The famous article by Garret Hardin (1968) on the tragedy of the commons suggests that the common pooling of resources leads to a labyrinth in which population growth, absence of assigned responsibilities and pursuit of profit end up exhausting natural resources, thus accelerating entropic processes that affect society and the environment. Hardin's model suggests a prisoner's-dilemma-type game in which competition among those who have natural resources leads to exhaustion or extinction, be it common pastureland for cattle, forests, fishing banks or water for crop irrigation. Since the article's publication, Hardin's model has had profound implications for conservationist policies. It presents sceptically the collaboration strategies that can fail due to the "desertion" of actors who refuse all communication/consensus with the goal of increasing their advantage over other actors, taking us back to the failures of collective action due to the presence of the free rider. Acceleration of entropic processes leads to the conclusion that systems must be governed by a central authority (the State) or by private entities. For Hardin, although State management is acceptable, private entities would be the real agents for actively diminishing the growth of entropy, because to generate a growing profit in economic units they are forced to guarantee resource sustainability before common use. Of course, this model does not consider whether the rationality of private agents in profit growth can lead to other behaviours, such as migrating or looking for new resources to exploit.

On the prescriptive models derived from Hardin's position of imposing a vertical and external coordination by the State or the market for the management of local actors and natural resources, Elinor Ostrom finds empirical signs that it is possible to develop alternative models based on a cooperation strategy, by drawing up binding contracts and conventions between actors. She bases this on an institutional approach to analyzing the constitution of complex systems understood as self-organization and self-management processes in the case of CPR. These in turn are understood as a system of natural or man-made resources that imply a very high cost if the idea is to exclude the potential beneficiaries (Ostrom 2000). However, there is always a risk of exclusion and conflict when extracted resource units (tons of fish, cubic meters of water, etc.) exceed the reproductive capacity of natural resource systems. Resource sustainability and conservation through time are among the priority goals of governance.

The capacity of institutions, that is, of the reference systems that serve as framework for decision-making, is put at risk in order to deal with situations of uncertainty. Often, institutions in Latin America are not necessarily supported by a conventional democratic political system, but by the performance of local societies. Family ties, social capital, implicit solidarity and reciprocity norms, local governments elected by custom and practice, informal credit and deep-seated modes of consumption that determine the valuation of the characteristics of a specific product are all elements that the different groups of producers have used to resist the crises caused by liberalization policies. Local institutional actions, however, are not integrated into or coordinated with public management to the same degree on other levels of government. In these cases, governance is based on a collective action with rural actors that invigorate and give cohesion to families and communities, so that if there is no continuity or sustainability in the action, success is not guaranteed. Non-sustainable leaderships largely provide an explanation for what Ostrom was obliged to acknowledge that she has no theoretical explanation for—why some collective actions are successful and some are not.

Ostrom's contributions allow us to tackle two other issues that are important for SYAL structure. The first concerns Amartya Sen's theory of capabilities (1985), which, although it refers to a person's achievements and functioning, also emphasizes self-esteem and the capability to insert oneself socially. Capabilities are not foreign to the formation of social abilities (Fligstein 2001) or to leadership formation. The second is the activation of territories through social networks (Gómez et al. 2006). Capabilities have a strong historic content that is transmitted through collective teaching and learning processes. Sen's outline and Ostrom's expansive social capital framework suggest that the resilience of marginalized communities can be increased for them to overcome conditions of poverty; thus, public policies aimed at promoting social skills, capabilities, trust and solidarity can be an important tool for leadership and SYAL sustainability.

This relates to the issue of territorial activation. In spite of discrepancies that could arise in terms of the concept of social capital, one of its characteristics is to facilitate social relations that permit the transmission of knowledge and the setting of common goals. Social capital can have different shapes: it can be open or closed; it can be functional to the interests of established groups or linked to the formation of civic and participatory cultures. Either way, social capital is a mesh that, in given territorial settings, allows collective know-how and behaviours to be transmitted. Social capital strengthens the operation of organizations and allows long-term commitments to be made. Social capital often takes the shape of self-regulated associations with strict criteria whose goal is to meet community needs. This last example is probably clearest in closed associations—such as mafias—, but can also be found in indigenous communities or those of indigenous origin, such as the community social capital found by Durston (2000) in some Latin American countries. Nevertheless, the attributes of community social capital can be transformed in terms of outside pressure. Under the forms of vertical coordination established by the global market, traditional institutions are maybe being blurred by migration, the demands of mass consumption and the presence of local leaderships and governments that have no interest in territorial development.

Some studies have in fact revealed this lack of interest, which is a great obstacle to collective action. Boucher mentions sixteen projects in the Lacandon jungle, in Mexico, which suffers from several coordination and organization flaws, due to a scant provision of public commodities, but also to the recent immigration of indigenous communities from the Chiapas highland to the Lacandon jungle. This implies serious difficulties in achieving a common institutional base and effective leadership. Another case of rupture in community social capital is the production and industrialization of nopal in the Tlanepantla municipality, in Morelos state, Mexico. This breaking off is accelerated by three factors that directly influence institutional changes in that municipality: the productive reconversion of the territory; the growing monetization of social life and the political break-off in the community, which has divided it into two opposing factions, each with its own organizations aimed at the market, but with a different degree of development. The institutional change that the municipality is going through has resulted in weak control and self-management of common resources, such as water distribution and the management of nearby forests that belong to the natural protected area of the Chichinautzin Biological Corridor. The damage to the community social capital has increased because the Federal Government has established conservationist policies in the area without effectively integrating social cooperation, so that many commoners have abandoned their responsibility in caring for the forest, leaving the task up to the government (Paz Salinas 2005). The adoption, twenty years ago, of nopal farming by municipal producers and the monetization of the local economy as a result of its proximity to the powerful consumption centre of Mexico City (42 kilometers away) have led to the abandonment of the milpa-based farming system. In other words, to the abandonment of polyculture—which offers the possibility of food security—as well as the political and productive breaking of the agreements that had supported community life until 2004. In that year the rupture between two groups of producers and subsequent armed intervention by the state and federal governments led to a considerable delay in plans to industrially transform the product and be innovative, with serious consequences for the community's social cohesion.

As an example of the development of the dominating agri-food model, a textbook case is the soy boom in Argentina, where two decades have brought about such an explosion that today approximately half of the country's farmable land is planted with soy (Elverdín et al. 2010). This development is based on a technological package called STSD (genetically modified soy through direct sowing) with grain that was genetically modified by adding the RR (Roundup Ready®) gene, which provides resistance to the non-selective glyphosate-based weed-killer sprayed over crops from light airplanes. On the other hand, direct sowing significantly reduces time and labour. The presence of large investors—often alien to the rural world—and the considerable trade volumes that favour the exclusion of local agents are some of the characteristics of this technologically and economic high-performance agri-food industry. This kind of model is mainstream, not only in Argentina, but internationally. Its development, however, poses serious ecologic and environmental problems: (i) consequences on biodiversity through the mass use of glyphosate-based weed-killers, (ii) water-table pollution, (iii) decrease in soil fertility through lack of crop rotation, (iv) wider deforestation. The very life of territories and the rural world is profoundly altered by this process which is leading to: (i) decreasing number of families living in rural areas, (ii) depopulation and disappearance of public infrastructure (schools, post offices, sanitation services), (iii) migration of professionals and loss of trade in rural areas. But at the same time, in a "sea of soy" many producers continue to make typical cheeses and sausages and to raise cattle in fields, because that is their trade and that is the Argentine meat they wish to produce. Among other things, examining SYAL starting from governance may facilitate an understanding of the current "rural world", and the antagonisms and complementarities between differing agri-food models, which is necessary for the creation of middle- and long-term public policies.

The problems mentioned so far are but a few examples of the difficulties posed by governance as a concept and a policy. To sum up, however, it is essential to point out that, given the new systems and forms of social action centred around food itself, it is necessary to make a theoretical effort that will never be sufficient because of the complexity and uncertainty of the observed phenomena. Facing this emergent complexity of systems, a heuristic method based on a concept network such as governance can be a good alternative for consolidating the scientific status of the SYAL approach and analytic category. But in the context of Latin America, it is necessary to reflect upon this category with certain specificities, such as a leadership structure based on property changes, the rise of new subjects, and the biodiverse, multiethnic and multicultural nature of societies. Likewise, it is necessary to view the asymmetric integration of rural agricultural industries due to their different abilities to face demand by markets in urban, peri-urban, regional and other levels and, finally, to face environmental risks that provide a new perspective of biodiversity conservation. All this reinforces the need to focus the structuring of the SYAL analytic category starting from the concept of governance.

3. Debate and conclusions. From the SYAL category to public policies. The traps of governance's double paradox

Collective action is expressed in different manners of territorial activation, such as appropriation of territorial resources or reaction to companies' externalities or to public policies that under an economic perspective of cluster formation are presented as agents of development to communities that are presumed to require development. They may incur a strong risk of breaking internal agreements and endogenous processes for institutional change. There is a need to obtain different certification and acknowledgment mechanisms to be present in urban-market niches or in differentiated niche markets in central countries. This implies, as pointed out by Muterbaugh and Lyon (2010), a series of self-exploitation and self-exclusion mechanisms by resource appropriators who, paradoxically, move within the frameworks of organic, certified or fair trade imposed by consumers. Faced with the inability to pay the high costs of certifiers, some organizations choose to specialize in certification tasks, but with high economic costs for the organization, creating conflict that opens huge rifts in social trust.

In other cases, externalities caused by enormous urban markets lead to the reconversion of farmland and the imposition of single-crop farming that question producers' technical knowledge and strategies. Nonetheless, this process implies assigning new meaning to action and leads to producing territorial identity seals for the adopted/appropriated resources in periods relatively short from a historic perspective. Numerous examples can be given in Latin America: nopal in the state of Morelos, the municipality of Tlanepantla, the main producer, has become known as "nopal sacred ground", and the cheeses and sausages of Tandil, in Argentina, a region that after several decades has become recognized as a producer of these typical products. These authenticities must be proven: to what degree do the image of a territory and its landscapes, as well as a product's taste and characteristics, contribute to building that very authenticity?

All of the above leads to an ethical assessment of SYAL from a biodiversity viewpoint. The problems of coordination, organization and leadership faced by collective action, and the adoption of individual strategies in the appropriation of resources such as water, soil and forests, have led to a lack of interest and solidarity in a series of production systems that have resulted in single crops, the excessive use of chemical fertilizers and careless management—euphemistically speaking—of fertilizer and pesticide containers. This has had consequences such as the silting of bodies of water, the pollution of surface waters and the loss of plant and animal species. The double paradox—that of mass consumption that imposes the need for distinction, and that of distinction, which carries a risk of exclusion—is also an ethical paradox that has been present since the beginning of some Latin American SYAL. The enhancement of products that are typical and identitarian and have territorial anchoring demanded by big cities or external markets is forcing processes of asymmetric integration onto markets. Geographical indications of origin, seals of quality, organic certification and other mechanisms of ethical commercial networks and local development public policies have introduced new forms of governance in which the paradox of a society of mass consumption and the reaction of the agents of differentiated consumption work in favour of the constitution of SYAL, but also in favour of their integration and repurposing in a non-sustainable exploitation scheme.

In spite of these problems, we do not believe that SYAL are a failed promise. The conclusions of our theoretical and empirical study of the development of SYAL are as follows. The governance of emerging systems depends in great measure not only on territorial localization of grouped companies that cooperate in production, services, tourism and gastronomy on a determined spatial scale in the manner of the Marshallian industrial district or clusters, but also on the acknowledgment of the sociological, ethical and political dimensions in the structuring of these systems. Amartya Sen's theory of capabilities and Elinor Ostrom's CPR open windows that are very important for recognizing that in an institutional view, it is possible to articulate collective-action efforts by building public policy alternatives to rural poverty. Nonetheless, this alternative should not be viewed from a productivist vantage point. We insist that institutional strengthening or construction be subject in this emerging governance to the environmental system and the conservation and innovation of natural—and, of course, cultural—wealth and heritage, and knowledge of biodiversity be applied.

The social capital that comes with SYAL structuring also needs to be defined in an expansive sense that implies the formation of institutions that foster and facilitate provision, responsibility and trust. As has been mentioned on different occasions, many of the problems faced by SYAL in Latin America are due to the weakness of institutional ties—which hinders market integration, innovation and the strengthening of collective strategies. This is related to a central element of collective action in Latin America: collective-action processes have not always gone hand in hand with the inclusion of territorial actors. Processes of innovation and of appropriation of a territory's resources have in fact often been in the hands of excluding rural elites and leaderships that foster the appearance and development of the free rider.

And the conclusions regarding mass consumption and differentiated consumption are clearly preliminary. There is the need to turn emerging systems into alternative systems in order to avoid the paradox to which they are subjected. A great deal of research is still necessary. It is also necessary to strengthen the methods for diagnosis and intervention in territories with potential productive systems that could become a SYAL in the sense of the triple specificity pointed out above: territories and institutions, technological innovation and food cultures. To date, methods have been of a qualitative nature and, in many ways, ethnographic. This gives us a precise idea of the territories, social actors, institutions and multiple functions of rural landscapes that make up tourism and the gastronomic culture, but there are too few statistics on frequencies and regularities to allow us to establish SYAL typologies. This is an issue for future studies. On the other hand, it is necessary to go beyond intervention strategies in strictly localized spheres, strategies that characteristically go hand in hand with a community's process of self-knowledge of its strengths, weaknesses, opportunities and threats (SWOT), to policies of promotion and facilitation of SYAL by governments at local, regional and other levels.

Lastly, despite acknowledgement of SYAL as a developing category, there are still no methods or techniques to determine the life cycle of SYAL. There is broad consensus that a flexible institutional base that understands territorial processes accompanies the initial structuring and strengthening of SYAL, but we have yet to discover under what conditions an emerging system can fail or end its life cycle. Sociological and very general hypotheses can be put forward, such as that failures in collective action and power structures in the rural sphere foster mistrust, false commitment and lack of ethics in the preservation of quality. This strengthens individual strategies and the breaking of ties between food and territory, except in some cases that are fortunately more common. Nevertheless, this hypothesis will need to be revised as we gain more knowledge on the matter.

References

Agamben, G. (1990). *La communauté qui vient. Théorie de la singularité quelconque*. Paris: Seuil. [*La comunità che viene*, Torino: Bollati Boringhieri, 2001]

Aguilar Villanueva, L.F. (2006). *Gobernanza y gestión pública*. Mexico: Fondo de Cultura Económica.

Álvarez Macías, A. et al. (eds.) (2006). *Agroindustria rural y territorio. Los desafíos de los sistemas agroalimentarios localizados*. Toluca, Mexico: Universidad Autónoma del Estado de México (UAEMEX).

Augé, M. (2005). *Los no lugares. Espacios del anonimato. Una antropología de la sobremodernidad*. Series: CLA-DE-MA. Antropología. [*Non-lieux. Introduction á une anthropologie de la surmodernité*. Translated by M. Mizraji]. Barcelona: Gedisa.

Beck, U. (2006). *La sociedad del riesgo. Hacia una nueva modernidad.* Surcos. [*Risikogesellschaft. Auf dem Weg in eine andere moderne.* Translated by J. Navarro, D. Jiménez and M. R. Borrás]. Barcelona, Buenos Aires, Mexico: Paidós.

Bengoa, J. (ed.) (2007). *Territorios rurales. Movimientos sociales y desarrollo territorial rural en América Latina.* Santiago de Chile: Catalonia.

Bordieu, P. (2000). *Las estructuras sociales de la economía.* Buenos Aires: Manantial.

Boucher, F. (2006). *Agroindustria rural y sistemas agroalimentarios locales. Nuevos enfoques de desarrollo territorial.* SIAL Network III International Congress. Food and Territory. Baeza, Spain: CSIC-Gis-SYAL-Agropolis. Retrieved from http://gis-syal.agropolis.fr/ALTER06/pdf/actes/boucher.pdf

Boucher, F. et al. (2000). *Globalización y evolución de la agroindustria rural en América Latina: Sistemas agroalimentarios localizados. Documentos de Trabajo Series.* Lima, Peru: PRODAR-IICA.

Brillat Savarin, J.A. (1825). *La physiologie du goût.* Paris: Hermann.

Cascante, M. (2002). *Sistemas agroalimentarios: Productos, empresas y dinámicas locales.* Montpellier, France: Ciat.

Durston, J. (2000). *¿Qué es el capital social comunitario?* Santiago de Chile: CEPAL (Social Development Division).

Elverdín, J. et al. (2010). Facteurs d'ancrage territorial des activités agricoles. Soja et élevage en Argentine. *Économie Rurale* 322: 64-78.

Filippi, M. and Torre, A. (2002). *Organisations et institutions locales. Comment activer la proximité géographic par des projects collectifs?* Montpellier, France: GIS-SYAL-INRA-CIRAD.

Fligstein, N. (2001). Social skill and the theory of fields. *Sociological Theory* 19 (2): 105-125.

Foucault, M. (1969). *L'archéologie du savoir.* Paris: Gallimard.

Gómez, C.A.C. et al. (2006). ¿Cómo activar los sistemas agroalimentarios localizados en América Latina? Un análisis comparativo. *Agroalimentaria* 22: 17-27.

Hardin, G. (1968). The tragedy of the commons. *Science* 162 (5364): 1243-1248.

Muchnik, J. (ed.) (1996). *Systèmes agroalimentaires localisés. (Organisations, innovations et développement local).* France: Département des systèmes agroalimentaires et ruraux, CIRAD, SAR. Retrieved from: http://www.gis-syal.agropolis.fr/index.php?option=com_content&view=article&id=41:premiere-notion-de-syal-en-1996.

Muchnik, J. et al. (2008). Systèmes agroalimentaires localisés. État des recherches et perspectives. *Cahiers Agricultures* 17 (6): 513-519.

Mutersbaugh, T. and Lyon, S. (2010). Transparency and democracy in certified ethical commodity networks. *Geoforum* 41 (1): 27-32.

Ortigues, E. (1989). Entretiens. *Le Coq-héron* 115: 58-72.

Ostrom, E. (2000). *El gobierno de los bienes comunes. La evolución de las instituciones de acción collectiva*. Sección de obras de economía contemporánea. [*Governing the commons. The evolution of institutions for collective action*. Translated by C. de Iturbide Calvo and A. Sandoval]. Mexico: Universidad Nacional Autónoma de México-Fondo de Cultura Económica.

Oyarzún, M.T. (2005). *Sellos de calidad en productos alimenticios agroindustriales con perspectiva para la pequeña agroindustria rural en América Latina*. In Cavalerie, A.M. (ed.), Seguridad Alimentaria. Cuadernos Fodepal. Biblioteca Virtual. Santiago de Chile: FODEPAL. Retrieved from: http://www.fodepal.org.

Paz Salinas, M.F. (2005). *La participación en el manejo de áreas naturales protegidas. Actores y intereses en conflicto en el corredor biológico Chichinautzin, Morelos*. Cuernavaca, Mexico: Universidad Nacional Autónoma de México.

Pecqueur, B. (2004). Vers une géographie économique et culturelle autour de la notion de territoire. *Géographie et Culture* 49: 71-86.

Polanyi, K. (2006). *La gran transformación. Los orígenes políticos y económicos de nuestro tiempo*. Economía. [*The great transformation. The political and economic origins of our time*. Translated by E.L. Suárez and R. Rubio]. Mexico: Fondo de Cultura Económica (FCE).

Poméon, T. et al. (2006). Las dinámicas colectivas en dos cuencas lecheras mexicanas: Tlaxco, Tlaxcala y Tizayuca, Hidalgo. *Agroalimentaria* 11 (22): 49-64.

Requier-Desjardins, D. (2001). Agroindustria rural y sistemas agroalimentarios localizados. ¿Cuáles apuestas? (Chap. Material de apoyo para el módulo 3.3, *En la promoción de la agroempresa rural para el desarrollo microregional sostenible*). *Memorias del II Curso Internacional*. Bogotá: International Center for Tropical Agriculture (CIAT).

Requier-Desjardins, D. (2004). *Produced common pool resources, collective action and sustainable local development. The case of food-processing clusters*. Versailles, France: Cahiers du C3ED. Retrieved from: http://www.scientificcommons.org/1544651.

Requier-Desjardins, D. (2006). Agroindustria rural, acción colectiva y SIALES: ¿desarrollo o lucha contra la pobreza? In Álvarez Macías, A.

et al. (eds.), *Agroindustria rural y territorio. Los desafíos de los sistemas agroalimentarios localizados*, 3-30. Mexico: Universidad Autónoma del Estado de México (UAEM).

Rodríguez-Borray, G. and Requier-Desjardins, D. (2006). La multifuncionalidad de los sistemas agroalimentarios locales en zonas rurales en países en desarrollo. El caso de la industria panelera colombiana. In Álvarez Macías, A. et al. (eds.), *Agroindustria Rural y Territorio,* 55-78. Mexico: Universidad Autónoma del Estado de Morelos (UAEM).

Sen, A. (1985). *Commodities and capabilities*. Amsterdam: Elsevier Science Publishing Company.

Torres Salcido, G. (2009). Los sistemas agroalimentarios locales. Capital social y instituciones. Una aproximación a la literatura reciente. In Romero, M.E. and Valverde Viesca, K. (eds.), *Teorías y problemas contemporáneos. Reflexiones desde la ciencia política*, 153-172. Colima, Mexico: Universidad de Colima, UNAM (Facultad de Ciencias Políticas y Sociales).

Torres Salcido, G. and Sanz Cañada, J. (2007). Construcción de capital social y sistemas agroalimentarios locales en áreas peri-urbanas de la Ciudad de México. In Méndez, R. (ed.), *Economías, mercados de trabajo y territorios metropolitanos en transformación*, 297-326. Madrid: Arethuse.

Von Chamisso, A. (1994). *La maravillosa historia de Peter Schlemihl*. Las tres edades. [*Peter Schlemihls wundersame Geschichte*. Translated by U. Michael and H. Valdés]. Madrid: Siruela.

CHAPTER SIX

DETERMINANTS OF INTER-ORGANIZATIONAL NETWORK PERFORMANCE—A LITERATURE REVIEW

VIRGINIE M. LEFEBVRE, ADRIENN MOLNÁR AND XAVIER GELLYNCK

Abstract

The aim of this chapter is to examine the literature addressing network performance and the factors influencing it. 29 papers published in 18 different journals were systematically reviewed through a multidimensional classification framework based on the strategy-structure-performance paradigm of strategic management. The outputs of the literature review reveal that inter-organizational networks are generally investigated for their impact on the performance of the firms participating in the networks. The performance of inter-organizational networks themselves is rarely investigated. Further, the review shows that the network structure can be characterized by five components i.e. network configuration, membership, tie, management and governance, which are all recognized as determinants of network performance. The nature of the link between these determinants and performance is however difficult to determine as performance is the result of the combination of network structure with the firm strategy towards the network and the external environment of the firm.

1. Introduction and purpose

During the last decades, various scholars in the field of strategic management and entrepreneurship have recognized the central role that inter-organizational networks play in firm competitiveness and growth (e.g., Stuart et al. 1999; Hite and Hesterley 2001; Westerlund et al. 2008). In a business world where knowledge is increasingly specialized and distributed across organizations, networks are now regarded as key vehicles for firms for obtaining access to new technologies and know-how

(Powell et al. 1996). Besides the access to new information beyond firm boundaries, vital for developing innovations and hence sustainable economic growth, the motives for firms for participating in networks are access to new resources, reduced transaction costs, economies of scale and risk or uncertainty sharing among network organizations (Hallikas et al. 2004; Kaler et al. 2000). In this perspective, having access to well-performing networks is of utmost importance for firms. Inter-organizational networks become particularly essential for small and medium enterprises (SME) especially regarding the access to new information and resources, as they often lack the necessary resources and capabilities to successfully innovate by sole means of in-house activities (Hausman 2005; Narula 2004; Nooteboom 1994). However, it is recognized that not all inter-organizational networks are equally performing (e.g., Dyer and Nobeoka 2000). Numerous studies have therefore started to examine performance of inter-organizational networks and the factors influencing it (e.g., Beamon 1998).

The purpose of this chapter is to examine the literature addressing the performance of inter-organizational networks and the factors influencing it. We believe that this is the first step towards the development of an adequate network performance measurement system which would tell how the network is performing, highlight possible improvements and help to diagnose problems (Cohen and Roussel 2005). In this chapter, inter-organizational networks are defined as long-term cooperative and purposeful relationships between firms and suppliers, customers, competitors, and other organizational actors in which organizations retain control over their own resources but jointly decide on their use (Brass et al. 2004; Möller and Svahn 2003). Following this definition, examples of inter-organizational network forms are strategic alliances, R&D consortia, buyer-supplier relationships and clusters or industrial districts (Gulati et al. 2000; Inkpen and Tsang 2005).

The chapter is structured as follow. Section 2 presents the research design of our study. In Section 3, the results of our literature review are presented. Finally, Section 4 highlights the main issues related to our findings and identifies the missing themes and gives direction for future research.

2. Research design

2.1 Literature review

The aim of this literature review is not to make a systematic review on what has been published on network performance; rather it aims at contributing to theory-building. In order to select the papers and to include as many "schools" as possible, we used Web of Science and Google Scholar as search engines. We made a systematic search for articles published between 1998 and 2009 which included the term "network" or "inter-organizational relationships" in the title, key words and/or abstract. Based on these criteria of selection, a total of 29 relevant articles were found in the following eighteen different journals (Table 6-1). Most of them were published in Strategic Management Journal and in Academy Management Review.

Table 6-1: Journals in which the articles selected for the study are published

Peer reviewed journals with impact factors	Peer reviewed journals without impact factors
Strategic Management Journal	The Academy of Management Journal
International Journal of Management Reviews	Journal of Business Research
American Journal of Agricultural Economics	Industrial Marketing Management
British Food Journal	Entrepreneurship and Regional Development
Journal of Public Administration Research and Theory	Management
The Academy of Management Review	Journal on Chain and Network Science
Journal of Management	Administrative Science Quarterly
Journal of International Management	
R&D Management	
Organization Science	
British Journal of Management	

The articles selected could be categorized in three different groups depending on the research approach: quantitative approach (48 percent of the papers), qualitative approach (7 percent of the papers) and conceptual approach including literature reviews (45 percent of the papers). Within the empirical papers, differences could be noted in terms of geographical region and industry covered (Table 6-2). Industries included for example the automotive, media, food processing, steel, semi-conductor and pharmaceutical biotechnology industries.

Table 6-2 Scope of the empirical papers

Country	Europe	USA/Canada	Asia	Cross-country	*Total*
Industry					
Mixed	3	4	1	1	9
Single	3	2	0	2	7
Total	6	6	1	3	16

Note: Oceania was considered in one paper (Blomstermo, Eriksson, Lindstrand, and Sharma 2004) which used cross-country data.

2.2 Multidimensional framework for classification of studies

Performance in general is known to be a complex concept, especially concerning its measurement. Authors in different disciplines have generally different views on what performance indicators should be included in the performance measurement system. As a consequence, a large number of performance indicators is used in literature; they are usually of different nature (i.e., qualitative vs. quantitative measures) and refer to different categories of measures (e.g., financial and operational measures) and different levels of performance (e.g., process, organization and supply chain levels) (Beamon 1999; Aramyan et al. 2007; Fabbe-Costes and Jahre 2008). Moreover, the selected papers indicate that network performance is influenced by many factors but the relation between these factors and network performance does not seem to be straightforward either. Network performance seems to be itself influenced by a network of interrelated and complex factors.

In order to cope with the above mentioned aspects related to performance, we used the strategy-structure-performance paradigm (SSP) of strategic management (Figure 6-1) (Defee and Stank 2005) as the basis of our multi-dimensional framework to classify the different papers (Table 6-3). The SSP paradigm was developed initially to explain the performance of a firm, but it has been extended since then to the supply chain. Since supply chains can be considered as being an inter-organizational network form (Omta 2002), the SSP paradigm which already has been extended to the supply chain, is applicable to networks as well. At the basis, the SSP paradigm predicts that a firm's strategy, created in consideration of external and internal (i.e., infrastructure) environmental factors, drives the development of organizational structure and processes, which in turn influence the firm's performance.

Figure 6-1 Strategy-structure-performance paradigm (Defee, and Stank 2005)

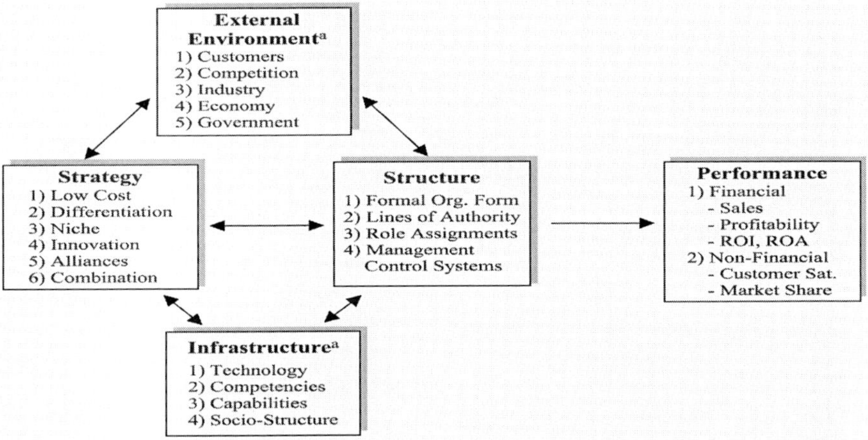

Note: [a]Contingent factors

Based on this paradigm, we analyzed the papers following eight dimensions (Table 6-3). First, as the focus of strategy are goals establishment and policies and plans development to achieve these goals (Ginsberg 1988; Andrews 1980), the papers were analyzed in relation to the consideration (or not) of goals and, where considered, regarding the type of goals e.g., innovation. As it recognized that firms develop relationships to achieve certain goals (e.g., Hallikas et al. 2004; Kale et al. 2000; Van de Ven 1976) and that inter-organizational networks are designed for specific purposes (e.g., Möller and Svahn 2003), we assume that goals that are set in a network can either refer to firms' goals, goals of the network as a whole or to dyad's goals, the dyad being the basic form of a relationship between two actors (Anderson et al. 1994). We therefore also distinguished between the papers depending on which level i.e., firm, dyad or network, the goals were related to.

The second dimension refers to the structure component of the paradigm. Adapted from the definition of organizational structure from Dalton et al. (1980) *network structure can be seen as the anatomy of the network providing a foundation within which the network functions*. Based on this definition, the papers were classified depending on whether the network structure was considered (or not), and where it was, the papers were classified based on the type of structure elements under study.

The third dimension based on which we classified the papers refers to network performance. Due to the wide variety of performance measures used in literature and the lack of clear and non-overlapping performance measures categories (Beamon 1998; Cohen and Roussel 2005; Beamon 1999; Stank et al. 2001), we chose, based on the premise that performance is defined as the level of achievement of goals (Kaplan 1983), to classify the papers based on whether goals with regard to performance were taken into account (or not) and on the type of goals. Furthermore, similarly to the strategy dimension of the framework, we distinguished the papers based on the level at which the performance was studied i.e., firm, dyadic, and network levels.

The fourth dimension relates to the contingent or environmental factors. Papers were classified depending on whether external and internal environmental factors of the network were considered or not, and if they were, on the type of contingent factors considered.

The last four dimensions are used to classify the papers related to the relationships, if any, between the different components of the paradigm. The fifth relational dimension refers to the relationship established between the contingent factors and the strategy; the sixth to the relationship established between the strategy and the network structure; the seventh to the relationship established between the contingent factors and the network structure; while the eighth and final dimension refers to the relationship established between the network structure and performance. The aim of classifying the papers based on these four relational dimensions is to determine what components so far have been recognized in literature to influence network performance.

Regarding all dimensions, papers were further classified based on whether the dimension considered was discussed or measured as the selected papers usually presented discrepancy in this regard. Table 6-3 summarizes the multidimensional classification framework used in this study.

Table 6-3 Multidimensional classification framework

Dimension	Coded element	Discussion	Measure
STRATEGY	Type of goal Goal level: - Firm level (F) - Dyadic level (D) - Network level (N)	Yes if discussed in the paper No if not discussed If yes, - type of goals discussed - three possible codes (i.e. F, D, N) depending on the level considered	Yes if measured in the paper No, if not measured If yes, - type of goals measured - three possible codes (i.e. F, D, N) depending on the level considered
STRUCTURE		Yes if discussed No if not discussed If yes, - type of structural elements discussed	Yes if measured No, if not measured If yes, - type of structural elements measured
PERFORMANCE	Goal based Type of goal Goal level: - F - D - N	Yes/G if performance is discussed in consideration of goals Yes/NG if it is discussed but not in consideration of goals No if it is not discussed If yes/G, - type of goals discussed - three possible codes (i.e. F, D, N) depending on the level considered	Yes/G if performance is measured in consideration of goals Yes/NG if it is measured but not in consideration of goals No, if it is not measured If yes/G, - type of goals discussed - three possible codes (i.e. F, D, N) depending on the level considered
CONTINGENT FACTORS (CF)		Yes if it discussed No if it is not discussed If yes, - type of contingent factors	Yes if it is measured No, if it is not measured If yes, - type of contingent factors
LINK BETWEEN CF & STRATEGY LINK BETWEEN STRATEGY & STRUCTURE LINK BETWEEN CF & STRUCTURE LINK BETWEEN STRUCTURE AND PERFORMANCE		Yes if it discussed No if it is not discussed If yes, - nature of the link	Yes if it is measured No, if it is not measured If yes, - nature of the link

3. Analysis

3.1 Discussing and measuring network strategy

About one third of the papers discuss to a certain extent goals, but none of the papers measure goals. Amongst the papers where goals are discussed, 6 refer to the firm's goals exclusively (Omta 2002; Fortuin and Omta 2009; Pittaway et al. 2004; Rowley et al. 2000; Goerzen and Beamish 2005; Koka and Prescott 2008), 2 to network's goals (Inkpen and Tsang 2005; Huggins 2000), 2 to both firm's goals and network's goals (Provan and Kenis 2007; van der Vorst 2006) and 1 to firm's-, network's- and individual goals i.e., goals of each person involved in the collaboration (Pitsis et al. 2004).

None of the papers have goals as direct focus, hence goals are not discussed extensively; but there are different types of goals or at least different types of competitive strategies which are referred to. Firstly, concerning goals at the firm level, it is stated by Goerzen and Beamish (2005) that it is widely recognized that the ultimate goals firms seek to reach are of an economic nature. In other words, firms seek to make profit. To make profit, firms try to increase their competitiveness via different competitive strategies. Based on the papers, some firms develop an innovation strategy which can be of three types. First, firms can follow a *prospector or explorer strategy* and focus on searching and experimenting to find innovations that will produce future profits (Fortuin and Omta 2009; Rowley et al. 2000; Koka and Prescott 2008). Second, they can develop a *defender or exploiter strategy* and are focus on searching for and exploiting existing technologies, skills and information (Pittaway et al. 2004; Rowley et al. 2000; Koka and Prescott 2008). Third, firms can opt for an *analyzer strategy*, situated somewhat in between the explorer and defender strategies (Koka and Prescott 2008). In addition, in Pittaway, Robertson et al. (2004), a "power strategy" is also mentioned as being used by firms looking for power over its suppliers or buyers for example. Secondly, there are a few network goals which are mentioned in the articles. In the public sector, network goals might for example include developing new clients, attracting network-funding, addressing community needs or improved client service (Provan and Kenis 2007) while in a supply chain network context, they might include differentiation, integrated quality and cost reduction (van der Vorst 2006).

3.2 Discussing and measuring network structure

Overall, the network structure is discussed in 40 percent of the papers and measured (to a certain extent) in 60 percent of the papers. Several structure elements are sometimes considered in the same paper, some of which are discussed and others measured. Based on the extension of the concepts of "structural" and "structuring" characteristics of organization of Campbell et al. to the network (Dalton et al. 1980), we classified the structure elements into two groups. The network structure elements were classified as either elements of the *structural dimension of network* referring to the physical characteristics of the network or as elements of the *structuring dimension of network* relating to the policies and activities occurring within the network that prescribe or restrict the behavior of network members (Table 6-4).

Based on the selected papers, there are three broad types of structure elements related to the structural dimension of network: network configuration, network membership and network ties. These are discussed in the subsequent paragraphs.

Network Configuration

Network configuration relates to the pattern of linkages among network members. There are four components of network configuration which are discussed and/or measured in the papers. The first component, referring to the weak connectedness between the network members of a firm, is *network sparseness* which is also known as the structural hole form of social capital. It is discussed in 5 papers (Gulati et al. 2000; Inkpen and Tsang 2005; Pittaway et al. 2004; van der Vorst 2006; Sporleder and Moss 2002) and measured in 5 papers (Rowley et al. 2000; Koka and Prescott 2008; Rodan and Galunic 2004; Zaheer and Bell 2005; Soda et al. 2004). The second component of network configuration is *network density* (also known as network closure or closure form of social capital), which refers, in contrast to network sparseness, to the high connectedness between the network members of a firm. Eight papers discuss network density to a certain extent (Gulati et al. 2000; Inkpen and Tsang 2005; Omta 2002; Pittaway et al. 2004; van der Vorst 2006; Sporleder and Moss 2002; Borgatti and Foster 2003; Hagedoorn et al. 2006) while 3 measure it (Rowley et al. 2000; Zaheer and Bell 2005; Soda et al. 2004). The third component of network configuration concerns the *pattern of direct and indirect ties*. Direct ties are ties that an organization has with its network partners while indirect ties are ties that an organization has with the partners of its network partners. Patterns of direct and indirect ties are discussed in Gulati et al. (2000) and measured in Beckman and Haunschild (2002). The fourth and final component of network configuration is *centrality* which refers to the extent to which a network revolves around one single firm in the network. Centrality is discussed in 8 articles (Gulati et al. 2000; Inkpen and Tsang 2005; Omta 2002; Pittaway et al. 2004; Pitsis et al. 2004; Rodan and Galunic 2004; Hagedoorn et al. 2006) and measured in 2 (Koka and Prescott 2008; Tsai 2001).

Network Membership

Network membership, referring to the composition of the network, is considered in the selected articles in terms of two components. First, network membership is discussed in 3 articles (Huggins 2000; Provan and Kenis 2007; Wincent et al. 2009) and measured in 2 articles (Goerzen and Beamish 2005; Ritter and Gemünden 2003) in terms of the *number of network members*. Second, it is discussed and/or measured in terms of the *type of network members* which relates to the size (Beckman and Haunschild 2002), industry (Gulati et al. 2000; Goerzen and Beamish 2005; Beckman and Haunschild 2002), scope of activities e.g., suppliers, customers (Gulati et al. 2000; Pittaway et al. 2004), resources (Gulati et al. 2000; Dyer and Singh 1998), past experiences and knowledge (Rodan and Galunic 2004; Beckman and Haunschild 2002), geographical situation (Gulati et. al. 2000; Huggins 2000) and innovativeness (Zaheer and Bell 2005).

Network Tie

Network tie refers to the characteristics of the relationships between network members. There are two components discussed and/or measured in the articles. First, authors have been interested in the *strength of ties* which refers, following Granovetter (1973), to the amount of time, emotional intensity, intimacy and reciprocal services which characterize the tie. Weak ties characterized by a low frequency and intensity of interactions are discussed in 5 papers (Dyer and Nobeoka 2000; Gulati et al. 2000; Inkpen and Tsang 2005; Pittaway et al. 2004; Thorpe et al. 2005) and measured in Rowley et al. (2000). Strong ties differentiated by a high frequency and intensity of interactions embedded of trust are also discussed in 5 papers (Dyer and Nobeoka 2000; Gulati et al. 2000; Inkpen and Tsang 2005; Sporleder and Moss 2002; Thorpe et al. 2005) and measured in Rowley, Behrens et al. (2000). The second component of network tie investigated by certain authors is *network multiplexity* which refers to the multiple relationships a firm develops with the same network member. Network multiplexity is measured in Beckman and Haunschild (2002). Besides the structural dimension of networks, papers investigate two different elements related to the structuring dimension of network: network management and network governance. These are discussed in the following paragraphs.

Network Management

Network management refers to the managerial and behavioral components that facilitate the execution of joint action across the whole network. Three components of network management are considered in the papers. The first refers to *conflict management*, i.e., the way conflicts between the network members are dealt with, and is measured in Kale, Singh et al. (2000). The second component of network management relates to *shared goals*, goal consensus or alignment of network-, firm- and (in some cases) individual level goals for network success. It is discussed in 5 papers (Inkpen and Tsang 2005; Huggins 2000; Provan and Kenis 2007; van der Vorst 2006; Pitsis et al. 2004). Finally, the third component refers to *network culture* and is discussed in Pitsis et al. (2004) and Inkpen and Tsang (2005). It should be mentioned that there is a fourth component of network management which is discussed in the selected papers and which concerns *building relational capital*. We decided to classify this component in the section concerning network governance as relational capital relates to informal governance mechanisms in most of the articles.

Network Governance

Network governance or tie modality involves the use of institutions and structures of authority and collaboration to direct, administrate, and control joint action across the whole network. There are three network governance components which are considered in the papers. The first component is *network board* defined by Wincent et al. (2009) as "*a group of individuals with supervisory power to make important decisions on major issues*". Network board is discussed in Provan and Kenis (2007) in terms of network board composition and independence i.e., network governed by the members themselves, by a single participating members or by outsiders. In Wincent et al. (2009), network board is measured in terms of board continuity relating to the rate at which board members are replaced, board size, board age, frequency of board meetings, board competence diversity and board independence. The second component of network governance discussed and/or measured in the papers relates to *formal network governance mechanisms*. These, discussed in 6 papers, mainly refer to the use of contractual arrangements (i.e., third-party enforcement of agreements) and equity arrangements or financial hostage (i.e., self-enforcing agreements) (Kale et al. 2000; Gulati et al. 2000; Omta 2002; Pittaway et al. 2004; Pitsis et al. 2004; Dyer and Singh 1998). Lastly, the third component of network governance considered by the authors refers to *informal network governance mechanisms*. These principally relate to reputation which is discussed in 4 papers (Kale et al. 2000; Gulati et al. 2000; Rowley et al. 2000; Dyer and Singh 1998) but also to relational capital (or social capital), which can be understood for the most part as the mutual trust that resides at the individual level between network partners. Relational capital or trust is discussed in 8 articles (Dyer and Nobeoka 2000; Gulati et al. 2000; Omta 2002; Pittaway et al. 2004; Huggins 2000; van der Vorst 2006; Pitsis et al. 2004; Dyer and Singh 1998) and measured in 1 (Kale et al. 2000).

Table 6-4 Network structure in the selected papers

Network dimension	Type of structure elements	Papers including only the discussed scope	Papers including the measured scope	Number of papers
STRUCTURAL	Network configuration			
	Network sparseness	Sporleder and Moss 2002; van der Vorst 2006; Pittaway et al. 2004; Gulati et al. 2000; Inkpen and Tsang 2005	Rodan and Galunic 2004; Rowley et al. 2000; Zaheer and Bell 2005; Soda et al. 2004; Koka and Prescott 2008	*10*
	Network density	Borgatti and Foster 2003; Omta 2002; Hagerdoon et al. 2006; Sporleder and Moss 2002; van der Vorst 2006; Pittaway et al. 2004; Gulati et al. 2000; Inkpen and Tsang 2005	Rowley et al. 2000; Zaheer and Bell 2005; Soda et al. 2004	*11*
	Pattern of direct and indirect ties	Gulati et al. 2000	Beckman and Haunschild 2002	*2*

Centrality	Pittaway et al. 2004; Pitsis et al. 2004; Omta 2002; Hagerdoon et al. 2006; Inkpen and Tsang 2005; Rodan and Galunic 2004; Gulati et al. 2000	Tsai 2001; Koka and Prescott 2008	*9*
Network membership			
Number of network members	Provan and Kenis 2007; Huggins 2000; Wincent et al. 2009	Goerzen and Beamish 2005	*4*
Type of network members:			
- *Size*		Beckman and Haunschild 2002	*1*
- *Industry*	Beckman and Haunschildt 2002; Gulati et al., 2000	Goerzen and Beamish 2005	*3*
- *Scope of activities*	Pittaway et al. 2004; Gulati et al. 2000		*2*
- *Resources*	Gulati et al. 2000; Dyer and Singh 1998		*2*
- *Past experience and knowledge*	Rodan and Galunic 2004; Beckman and Haunschildt 2002		*2*
- *Geographical situation*	Gulati et al. 2000; Huggins 2000		*2*
- *Innovativeness*		Zaheer and Bell 2005	*1*
Network tie			
Strength of ties:			
Weak ties	Pittaway et al. 2004; Gulati et al. 2000; Thorpe et al. 2005; Dyer and Nobeoka, 2000: Inkpen and Tsang 2005	Rowley et al. 2000	*6*
Strong ties	Sporleder and Moss 2002; Gulati et al. 2000; Thorpe et al. 2005; Dyer and Nobeoka 2000 ; Inkpen and Tsang 2005	Rowley et al. 2000	*6*
Network multiplexity		Beckman and Haunschild 2002	*1*

STRUCTURING	Network management			
	Conflict management		Kale et al. 2000	*1*
	Shared goals	Huggins 2000; Pitsis et al. 2004; Inkpen and Tsang 2005; Provan and Kenis 2007; van der Vorst 2006		*4*
	Network culture	Pitsis et al. 2004; Inkpen and Tsang 2005; Dyer and Singh 1998		*3*
	Network governance			
	Network board	Provan and Kennis 2007	Wincent et al. 2009	*2*
	FNG mechanisms	Pittaway et al 2004; Gulati et al. 2000; Kale et al. 2000; Pitsis et al. 2004; Omta 2002; Dyer and Singh 1998		*6*
	ING mechanisms:			
	- *Reputation*	Dyer and Singh 1998; Gulati, Nohria et al. 2000; Kale, Singh et al. 2000; Rowley, Behrens et al. 2000		*4*
	- *Relational capital*	Huggins 2000; Omta, 2002; Dyer and Singh 1998; Pitsis, Kornberger et al. 2004; Pittaway, Robertson et al. 2004; Gulati, Nohria et al. 2000; van der Vorst 2006; Dyer and Nobeoka 2000	Kale, Singh et al. 2000	*9*

Notes: FNG mechanisms, formal network governance mechanisms, ING mechanism, informal network governance mechanisms.

3.3 Discussing and measuring performance

Performance is discussed in consideration of goals in one paper (van der Vorst 2006) and measured in two (Fortuin, and Omta 2009; Goerzen, and Beamish 2005) (Table 6-5). In van der Vorst (2006), the focus is on the performance at the network level and both network and firm level goals are considered. It is stated that performance should be measured based on a number of performance metrics which take first into account the network goals and second the goals of each network member. In Fortuin and Omta (2009) and Goerzen and Beamish (2005), the focus is on the performance at the firm level and only the firm level goals are taken into account. These relate to innovation (Fortuin and Omta 2009) and profit (Fortuin and Omta 2009; Goerzen and Beamish 2005); hence in these articles performance refers either to innovation and/or economic performance.

Besides the papers with a clear link between performance and goals, there are a number of papers dealing with performance but not in consideration of goals (Table 6-5). Most of these papers (11 out of 18) focus on the performance at firm level (Kale et al. 2000; Dyer and Nobeoka 2000; Gulati et al. 2000; Pittawayet al. 2004; Rowley et al. 2000; Koka and Prescott 2008; Sporleder and Moss 2002; Zaheer and Bell 2005; Beckman and Haunschild 2002; Wincent et al. 2009; Ritter and Gemünden 2003) while five focus on the performance at the network level (Huggins 2000; Provan and Kenis 2007; Pitsis et al. 2004; Soda et al. 2004; Dyer and Singh 1998) and none on the performance at the dyad level. Interestingly, two papers investigate performance at levels others than the ones which have been considered in our framework: the intra-organizational business unit and individual levels (Rodan and Galunic 2004; Tsai 2001). It should be mentioned that Soda et al. (2004) do not explicitly refer to performance at the network level. Nevertheless, we decided to classify the article in the network level category as it focuses on the performance of TV productions which are the results of a number of specialists working together.

In the 18 papers dealing with performance but not in consideration of goals, performance is of seven different types. The most common types are *economic performance*, *innovation performance* and *learning performance*. Besides these, some articles discuss performance in terms of *success and failure* of networks which refer to the ability of networks to become a sustained and valued form of business activity for their members (Huggins 2000). Other articles discuss performance in terms of network *effectiveness* which relates to the attainment of outcomes which could not have normally been achieved by network member acting independently while still some other articles discuss performance in terms of *managerial performance*. In addition to these types of performance, there is another type of performance which we classify under "*others*", being rather particular and specific to the sector under investigation. This refers to the TV productions performance in Soda et al. (2004) which was measured *via* the size of the audience.

Table 6-5 Performance in the selected papers

Performance level measurement	Performance type	Papers including only the discussed scope		Papers including the measured scope		Number of papers
		Yes/G	*Yes/NG*	*Yes/G*	*Yes/NG*	
Network	Economic performance	van der Vorst 2006				*1*
	Innovation performance					
	Learning performance					
	Managerial performance					
	Success/failure		Huggins 2000			*1*
	Effectiveness		Provan and Kenis 2007; Dyer and Singh 1998; Pitsis et al. 2004			*3*
	Others				Soda et al. 2004	*1*
Dyadic	Economic performance					
	Innovation performance					
	Learning performance					
	Managerial performance					
	Success/failure					
	Effectiveness					
	Other					

Firm	Economic performance	Sporleder and Moss 2002; Gulati et al. 2000	Goerzen and Beamish 2005; Fortuin and Omta 2009	Rowley et al. 2000; Zaheer and Bell 2005; Koka and Prescott 2008	*7*
	Innovation performance	Pittaway et al. 2004	Fortuin and Omta, 2009	Wincent et al. 2009; Ritter and Gemünden 2003	*4*
	Learning performance			Beckman and Haunschild 2002; Dyer and Nobeoka 2000; Kale et al. 2000	*3*
	Managerial performance				
	Success/failure				
	Effectiveness				
	Others				
Intraorg. BU	Economic performance			Tsai 2001	*1*
	Innovation performance			Tsai 2001	*1*
	Learning performance				
	Managerial performance				
	Success/failure				
	Effectiveness				
	Others				
Individual	Economic performance				
	Innovation performance			Rodan and Galunic 2004	*1*
	Learning performance				
	Managerial performance			Rodan and Galunic 2004	*1*
	Success/failure				
	Effectiveness				
	Others				

Note: Intraorg. BU, intra-organizational business unit.

3.4 Discussing and measuring contingent factors of the network

As mentioned in the SSP paradigm, contingent factors refer either to internal environmental factors (or infrastructure) or to external factors. These will be discussed one after another in the next paragraphs.

Internal Environmental Factors

In the selected papers, the environmental factors, when considered, generally refer to the internal environmental factors of the firm participating in the network and not to the internal environmental factors of the network itself. In the following papers, internal factors of the firm determine the extent to which the firm will develop relationships with other organizations and refer to networking capability (Kale et al. 2000; Gulati et al. 2000; Goerzen and Beamish 2005; Hagedoorn et al. 2006; Ritter and Gemünden 2003; Möller and Halinen 1999; Ritter et al. 2004), innovation capability (Fortuin and Omta 2009; Zaheer and Bell 2005; Tsai 2001), network and net visioning capability (Möller and Halinen 1999), level of technical and commercial competence (Pittaway et al. 2004), firm resources (Fortuin and Omta 2009), organizational culture (Omta 2002; Goerzen and Beamish 2005; Pitsis et al. 2004) and network experiential knowledge (Blomstermo et al. 2004). There are however two papers (Huggins 2000; Pitsis et al. 2004) which consider the internal environmental factors of the network by discussing the *competencies of the network broker or leader* in generating valid interaction and exchange between participants and maintaining an equal distribution of power and dependency in order to alleviate the potential for conflicts to emerge.

External Environmental Factors

As for the internal environmental factors, external factors, when considered, generally refer to the external environmental factors of the firm participating in the network such as the competitive environment (Omta 2002; Fortuin and Omta 2009; Rowley et al. 2000; Koka and Prescott 2008; Sporleder and Moss 2002) and the social environment (Goerzen and Beamish 2005; Thorpe et al. 2005), and not to the external environmental factors of the network itself. There are however a few papers (Pittaway et al. 2004; Dyer and Singh 1998; Thorpe et al. 2005) where the external environmental factors of the network are considered. In these papers, the focus is on the *institutional and policy environment* which includes the legal system, banking and finance system, structure of the labor market and the education and political systems.

3.5 Discussing the links between the components of the SSP paradigm

Link between Contingent Factors of the Network and Network Strategy

No paper discusses or measures the link between contingent factors of the network and strategy of the network. In the selected papers, there are however three papers which discuss the link between the competitive environment, an external environmental factor of the firm, and the strategy of the firm (Rowley et al. 2000; Koka and Prescott 2008; Sporleder and Moss 2002). Following these papers, the choice of a firm to opt for a certain innovation strategy depends on the conditions of the environment, and especially on its uncertainty. Following e.g., Rowley et al. (2000), a firm in a stable environment will more likely develop an exploiter innovation strategy while in an unstable environment, the same firm will more opt for an explorer innovation strategy.

Link between Network Strategy and Network Structure

Only one paper (van der Vorst 2006) discusses the link between strategy of the network and network structure. In this article, network management, one of the structuring dimensions of the network, is directly related to the objectives of the network. Besides this paper, 4 other papers discuss the link between strategy of the firm and network structure, focusing exclusively on the structural dimension of the network (Pittaway et al. 2004; Rowley et al. 2000; Koka and Prescott 2008; Sporleder and Moss 2002). In these papers, the focus is on *network configuration*, *strength of ties* and *network membership* and on two types of competitive strategies, i.e. *innovation* and *"power" strategy*. Based on the papers, network configuration, strength of ties, and network membership is contingent on the firm strategy. Following e.g., Rowley et al. (2000) where innovation strategies are discussed, a firm having an exploiter strategy, and therefore interested in gaining specific know-how, will prefer a dense network over a sparse one. On the contrary, a firm having an explorer strategy will prefer a sparse network in order to have access to new and non-overlapping information. Moreover, in the context of a firm having an innovation strategy, the network configuration depends also, following Pittaway et al. (2004), on the type of innovation required, i.e., process vs. product innovation, while the type of network members is contingent on whether the firm seeks to develop incremental or radical innovation. In case of a firm having a "power strategy", it will seek for centrality in order to gain power over network members (Pittaway et al. 2004).

Link between Contingent factors of the Network and Structure

There are a few papers which discuss and/or measure the link between contingent factors and structure. We will first review the papers where internal environmental factors are considered and then papers where external factors are considered.

Internal factors

Two papers discusses the link between internal factors of the network and network structure (Huggins 2000; Pitsis et al. 2004). In these papers, the authors highlight the importance of the *competencies of the network broker or leader* in harnessing the different interests and attitudes of network members in a format that generates valid interaction; or in other words the authors highlight the importance of the competencies of the network broker in developing shared goals and a network culture. In Huggins (2000), the importance of the role of the broker in conflict management is also emphasized.

It should be mentioned that none of the selected papers measure the link between contingent factors of the network and network structure.

External factors

Two papers (Pittaway et al. 2004; Thorpe et al. 2005) discuss the link between external environmental factors of the network and network structure, but none measure it. The focus in these two papers is on the *institutional and policy environment*. Following the articles, the institutional and policy environment plays a role in facilitating the development of networks. Governments can shape the cultural conditions and infrastructure for networking but also can assist in brokering relationships and acting as intermediaries. Moreover, the institutional and policy environment also plays a role in shaping network configurations as it influences inclinations towards trust, legal contracting, opportunism and self-interest.

Link between Network Structure and Performance

Several papers discuss and/or measure the link between structure and performance. In the following paragraphs, we will examine one after another each of the network structural elements presented in Table 4, i.e., network configuration, network membership, network tie, network management and network governance. It should be mentioned that for the majority of the components of network structure, no general conclusion concerning the nature of the link will be drawn. Comparing the results of different articles on one particular component is, in most cases, not appropriate for two main reasons. First, articles might measure performance at different level e.g., network and firm level. Second, they might refer to different types of performance e.g., economic and learning performance.

Network configuration

There are 2 papers where the link between network configuration and performance is discussed (Gulati et al. 2000; Sporleder and Moss 2002) and 6 where the link is measured (Rowley et al. 2000; Koka and Prescott 2008; Rodan and Galunic 2004; Zaheer and Bell 2005; Soda et al. 2004; Tsai 2001), covering together the four different components of network configuration described in section III.B i.e., *network sparseness*, *network density*, *pattern of direct and indirect ties* and *centrality* (Table 6-6). It should be mentioned that Sporleder and Moss (2002) mainly refer to the results of Rowley et al. (2000) when they discuss network sparseness/network density and performance. According to the articles, there is no optimal network configuration but the benefits of a certain type of network configuration vary according to the situation. The impact of one type of network configuration on performance seems to depend first on the type of performance under investigation. As an example, a business unit's centrality in its intra-organizational network is recognized in Tsai (2001) as positively and significantly influencing innovation performance while it does not have any impact on the economic performance of the business unit. Second, the benefits of a certain network configuration on performance might be triggered by environmental changes. In e.g., Koka et al. (2008), it is shown that, following a legislative change, a firm centrally positioned in a network will exhibit lower economic performance while a firm occupying a bridging position in a sparse network higher economic performance. Third and last, it is not solely a particular network configuration which influences performance but rather a combination of this particular network configuration with a certain network strategy and/or other network characteristics such as network membership. As a first example, it is recognized in Koka et al. (2008) that a firm pursuing an analyzer strategy and a central position within the network will exhibit higher relative performance than firms pursuing a defender strategy. As a second example, Rodan and Galunic (2004) show that a sparse network will positively be linked to innovation performance if it is combined with a high level of knowledge heterogeneity within the network members.

Network membership

There are 7 articles dealing with the link between network membership and performance (Table 6-6); 4 of which discussing this link (Gulati et al.

2000; Pittaway et al. 2004; Huggins 2000; Dyer and Singh 1998) while 3 measuring it (Goerzen and Beamish 2005; Rodan and Galunic 2004; Beckman and Haunschild 2002). Both components of network membership of Table 6-4, i.e., *number of network members* and *type of network members*, are considered in the articles (Table 6-6). It should be mentioned that articles discuss the type of network members in terms of four aspects: diversity (i.e., how diverse network members are in terms e.g., of knowledge and type of industry covered), proximity (i.e., how geographically close are network members), complementarity (i.e., how complementary network members are in terms of e.g., resources and capabilities), and innovativeness (i.e., how innovative are network members). Following the articles, the aspect of network membership which has received the most attention is network diversity. Network diversity is discussed either in terms of diversity of scope of activities of network members (Pittaway et al. 2004), diversity of the nature of the businesses involved (Huggins 2000), number of unique relationships with the focal firm and number of unique industries of partners (Goerzen and Beamish 2005), past experiences (including good/bad and small/big experiences), patterns of direct and indirect ties and network members' size (Beckman, and Haunschild 2002), and heterogeneity of knowledge (Rodan, and Galunic 2004). It is interesting to notice that in the articles concerned with innovation and/or learning performance (Pittaway et al. 2004; Rodan and Galunic 2004; Beckman and Haunschild 2002), support is given to the idea that there is a positive link between network diversity and performance while the articles focusing on economic performance and/or network success and failure do not support such a link (Goerzen and Beamish 2005; Huggins 2000). They instead provide evidence that firms maintaining a homogenous network enjoy superior performance than firms maintaining a heterogeneous network. Regarding economic performance, network diversity seems only to become beneficial for firms beyond a certain point i.e., when firms develop a highly heterogeneous network instead of "an average heterogeneous" network (Goerzen and Beamish 2005). It should be mentioned that diverging results are also obtained for the nature of the link between the number of network members and performance depending on the type of performance considered. In Huggins (2000), where the focus is on the success and failure of networks, it is concluded that the higher the number of network members the poorer the performance while in Ritter and Gemünden (2003), where the focus is on the innovation performance of the firm, it is concluded the opposite.

Network tie

The link between network tie and performance is discussed in 1 paper (Gulati et al. 2000) and measured in 3 (Dyer and Nobeoka 2000; Rowley et al. 2000; Beckman and Haunschild 2002). It should be mentioned that Gulati et al. (2000) mainly refers to Dyer and Nobeoka (2000) when they discuss the strength of ties. Both components of network tie, i.e., *network strength* and *network multiplexity*, are considered in the papers, network strength being the one the most studied in the articles (Table 6-6). Following the articles, strength of ties has especially received attention in the context of learning and the benefits of strong/weak ties seem to be contingent on the environment. Strong ties, associated with trust and fine-grained information exchange between partners, lead to higher economic and learning performance than weak ties in stable environments where an exploitation strategy is used (Dyer and Nobeoka 2000; Rowley et al. 2000). On the contrary, weak ties, through which novel information is accessible, lead to higher economic performance than strong ties in uncertain environments where an exploration strategy is used (Rowley et al. 2000).

Network management

There are three articles which discuss (Pitsis et al. 2004; Dyer and Singh 1998) and measure (Kale et al. 2000) the link between network management and performance (Table 6-6). Four aspects are considered in the articles i.e., *conflict management*, *incentives to encourage transparency and discourage free riding, shared goals* and *network culture*. Hence all components of network management are explicitly considered. In Kale et al. (2000), where the focus is on conflict management, it is shown that integrative conflict management, i.e., joint management of conflict with mutual concern for "win-win" for all concerned, is positively linked to the learning performance of the firm. Following Dyer and Singh (1998), incentives to encourage transparency and discourage free riding are also positively linked to the learning performance of the firm while shared goals and strong network culture are positively linked to the success of networks in general (Pitsis et al. 2004).

Network governance

The link between network governance and performance is discussed in 4 articles (Huggins 2000; Provan and Kenis 2007; Pitsis et al. 2004; Dyer and Singh 1998) and measured in 2 (Kale et al. 2000; Wincent et al. 2009). All three components of network governance, i.e. *network board*, *formal network governance mechanisms* and *informal network governance mechanisms*, are considered (Table 6). There are two points worth noticing concerning the link between network governance and performance. The first concerns board independence. Wincent et al. (2009) support the idea that board independence is negatively related to performance while Provan and Kenis (2007) have a more moderate idea about the relationship between board independence and performance and suggest that the impact of board independence on performance is contingent on the level of trust between network members, the number of network members, the degree of goal consensus and the need for network-level competencies. The second point concerns network governance mechanisms. It seems that, whether performance refers to economic, learning or other types of performance, there is a consensus among authors that informal network governance mechanisms, especially relational capital, lead to higher performance than formal network governance mechanisms.

Table 6-6 Link between structure and performance

Type of structural elements	Aspect discussed	Papers including only the discussed scope				Papers including the measured scope			
		-	0	+	Context dependent	-	0	+	Context dependent
Network configuration									
Network sparseness				Gulati et al. 2000	Sporleder and Moss 2002			Rodan and Galunic 2004; Zaheer and Bell 2005	Rowley et al. 2000; Koka and Prescott 2008; Soda et al. 2004
Network density					Sporleder and Moss 2002		Zaheer and Bell 2005		Rowley et al. 2000; Soda et al. 2004
Pattern of direct and indirect ties				Gulati et al. 2000					
Centrality	Central position			Gulati et al. 2000			Tsai 2001		Koka and Prescott 2008
Network membership									
Number of network members	High number					Huggins 2000		Ritter and Gemünden 2003	

Type of network members	Diversity			Pittaway et al. 2004		Huggins 2000		Rodan and Galunic 2004; Beckman and Haunschildt 2002 Huggins 2000	Goerzen and Beamish 2005
	Proximity	Gulati et al. 2000							
	Complementarity			Dyer and Singh 1998					
	Innovativeness						Zaheer and Bell 2005		
Network tie *Strength of ties*	Strong ties			Gulati et al. 2000				Dyer and Nobeoka 2000 Beckman and Haunschild 2002	Rowley et al. 2000
Network multiplexity									
Network management	Conflict management Incentives to encourage transparency and discourage free riding			Dyer and Singh, 1998				Kale et al. 2000	

	Shared goals			Pitsis et al. 2004					
	Network culture			Pitsis et al. 2004					
Network governance									
Network board	Network board continuity								Wincent et al. 2009
	Large board							Wincent et al. 2009	
	Board competence diversity					Wincent et al. 2009		Wincent et al. 2009	
	High frequency of meetings			Provan and Kenis 2007					
	Board independence					Wincent et al. 2009			
FNG mechanisms		Dyer and Singh 1998							
ING mechanisms				Dyer and Singh, 1998; Pitsis et al. 2004; Huggins 2000				Kale et al. 2000	

Notes: FNG mechanisms, formal network governance mechanisms, ING mechanism, informal network governance mechanisms.

4. Conclusion

The aim of this chapter was to examine the literature addressing the performance of inter-organizational networks and the factors influencing it. Our review shows that network performance has especially received attention in literature in terms of the impact the inter-organizational network has on the performance of the firm participating in the network rather than in terms of the performance of the network itself. Based on Provan et al. (2007), this situation might be due to the fact that the performance of inter-organizational network at the network level appears to be less important than the performance of its members, these networks being rather bottom-line oriented and members' goal driven. Moreover, it appears from our review that, although performance relates to the level of achievement of goals (Kaplan 1983), few papers discuss or measure performance in consideration of goals. In addition, the outcomes of our review illustrate that in papers where goals are considered, goals generally refer to the goals of the firms participating in the network and not to the goals of the network itself. These findings lead to the question whether goals of inter-organizational do really exist on their own and whether the level of achievement of goals of the network is an appropriate criterion to measure network performance. Further research is required to answer this question.

Our analysis of the papers further shows that network structure is characterized by three structural components i.e., network configuration, membership and tie referring to the physical characteristics of the network and by two structuring components i.e., network management and governance relating to the policies and activities occurring within the network that prescribe or restrict the behaviour of network members. The influence of all these components on the performance of the firm participating in the network has been extensively investigated in literature. No general conclusion however can be drawn concerning the nature of the link between these components and performance, as from our review it appears that performance is the result of the combination of network structure with the firm strategy towards the network and the external environment of the firm. This was supported in different articles such as Koka and Prescott (2008) and Rowley et al. (2000) where it was shown that a particular network configuration had a positive impact on performance if combined with a certain network strategy and/or particular environment. Moreover, our review suggests that, in an equal context, i.e., same firm strategy towards the network and same external environment of the firm, different combinations of network structure elements lead to different network performance. As an example, Rodan and Galunic (2004) showed that a sparse network will be positively linked to innovation performance if it is combined with a high level of knowledge heterogeneity within the network members.

Finally, our review reveals two factors influencing the network structure, the competencies of the network broker and the institutional and policy environment. As only two contingent factors were recognized to influence the network structure, it would be interesting to extend this study to further investigate the contingent factors of networks and the relationships between them and other components of the paradigm i.e., network strategy and network structure.

References

Anderson, J. et al. (1994). Dyadic Business Relationships within a Business Network Context. *Journal of Marketing* 58: 1-5.

Andrews, K.R. (1980). *The Concept of Corporate Strategy*. Irwin: Homewood.

Aramyan, L.H. et al. (2007). Performance Measurement in Agri-Food Supply Chains: A Case Study. *Supply Chain management* 12 (4): 304-315.

Beamon, B.M. (1998). Supply Chain Design and Analysis: Models and Methods. *International Journal of Production Economics* 55 (3): 281-294.

—. (1999). Measuring Supply Chain Performance. *International Journal of Operations and Production Management* 19 (3): 275-292.

Beckman, C.M. and Haunschild, P.R. (2002). Network Learning: The Effects of Partners' Heterogeneity of Experience on Corporate Acquisitions. *Administrative Science Quarterly* 47 (1): 93-124.

Blomstermo, A. et al. (2004). The Perceived Uselfulness of Network Experiential Knowledge in the Internationalizing Firm. *Journal of International Management* 10: 355-373.

Borgatti, S.P. and Foster, P.C. (2003). The Network Paradigm in Organizational Research: A Review and Typology. *Journal of Management* 29 (6): 991-1013.

Brass, D.J. et al. (2004). Taking Stock of Networks and Organizations: A Multilevel Perspective. *Academy of Management Journal* 47 (6): 795-817.

Cohen, S. and Roussel, J. (2005). *Strategic Supply Chain Management. The Five Disciplines for Top Performance*. New York: McGraw-Hill.

Dalton, D.R. et al. (1980). Organization Structure and Performance: A Critical Review. *Academy of Management Review* 5 (1): 49-64.

Defee, C.C. and Stank, T.P. (2005). Applying the Strategy-Structure-Performance Paradigm to the Supply Chain Environment. *The International Journal of Logistics* 16 (1): 28-50.

Dyer, J.H. and Nobeoka, K. (2000). Creating and Managing a High-Performance Knowledge-Sharing Network: The Toyota Case. *Strategic Management Journal* 21: 345-367.

Dyer, J.H. and Singh, H. (1998). The Relational View: Cooperative Strategy and Sources of Interorganizational Competitive Advantage. *The Academy of Management Review* 23 (4): 660-679.

Fabbe-Costes, N. and Jahre, M. (2008). Supply Chain Integration and Performance: A Review of the Evidence. *The International Journal of Logistics Management* 19 (2): 130-154.

Fortuin, F.T.J.M. and Omta, O.S.W.F. (2009). Innovation Drivers and Barriers in Food Processing. *British Food Journal* 111 (8): 839-851.

Ginsberg, A. (1988). Measuring and Modelling Changes in Strategy: Theoretical Foundations and Empirical Directions. *Strategic Management Journal* 9 (6): 559-575

Goerzen, A. and Beamish, P.W. (2005). The Effect of Alliance Diversity on Multinational Enterprise Performance. *Strategic Management Journal* 26, 333-354.

Granovetter, M.S. (1973). The Strength of Weak Ties. *American Journal of Sociology* 78: 1360-1380.

Gulati, R. et al. (2000). Strategic Networks. *Strategic Management Journal* 21: 203-215.

Hagedoorn, J. et al. (2006). Inter-Firm R&D Networks: The Importance of Strategic Network Capabilities for High-Tech Partnership Formation. *British Journal of Management* 17: 39-53.

Hallikas, J. et al. (2004). Risk Management Processes in Supplier Networks. *International Journal Production Economics* 90: 47-58.

Hausman, A. (2005). Innovativness among Small Businesses: Theory and Propositions for Future Research. *Industrial Marketing Management* 34: 773-782.

Hite, J.M. and Hesterley, W.S. (2001). The Evolution of Firm Networks: From Emergence to Early Growth of the Firm. *Strategic Management Journal* 22: 275-286.

Huggins, R. (2000). The Success and Failure of Policy-Implanted Inter-Firm Network Initiatives: Motivations, Processes and Structure. *Entrepreneurship and Regional Development* 12: 111-135.

Inkpen, A.C. and Tsang, E.W.K. (2005). Social Capital, Networks, and Knowledge Transfer. *Academy of Management Review* 30 (1): 146-165.

Kale, P. et al. (2000). Learning and Protection of Proprietary Assets in Strategic Alliances: Building Relational Capital. *Strategic Management Journal* 21 (3): 217-237.

Kaplan, R.S. (1983). Measuring Manufacturing Performance: A New Challenge for Managerial Accounting Research. *Accounting Review* 58 (4): 686-721.

Koka, B.R. and Prescott, J.E. (2008). Designing Alliance Networks: The Influence of Network Position, Environmental Change, and Strategy on Firm Performance. *Strategic Management Journal* 29: 639-661.

Möller, K.K. and Halinen, A. (1999). Business Relationships and Networks. *Industrial Marketing Management* 28: 413-427.

Möller, K.K. and Svahn, S. (2003). Managing Strategic Nets: A Capability Perspective. *Marketing Theory* 3 (2): 201-226.

Narula, R. (2004). R&D Collaboration by Smes: New Opportunities and Limitations in the Face of Globalisation. *Technovation* 24 (2): 153-161.

Nooteboom, B. (1994). Innovation and Diffusion in Small Firms—Theory and Evidence. *Small Business Economics* 6 (5): 327-347.

Omta, O.S.W.F. (2002). Innovation in Chains and Networks. *Chain and Network Science* 2 (2): 73-80.

Pitsis, T.S. et al. (2004). The Art of Managing Relationships in Interorganizational Collaboration. *Management* 7 (3): 47-67.

Pittaway, L. et al. (2004). Networking and Innovation: A Systematic Review of the Evidence, *International Journal of Management Reviews* 5/6 (3&4): 137-168.

Powell, W.W. et al. (1996). Interorganizational Collaboration and the Locus of Innovation: Networks of Learning in Biotechnology, *Administrative Science Quarterly* 41: 116-145.

Provan, K.G. and Kenis, P. (2007). Modes of Network Governance: Structure, Management, and Effectiveness. *Journal of Public Administration Research and Theory* 18: 229-252.

Provan, K.G. et al. (2007). Interorganizational Networks at the Network Level: A Review of the Empirical Literature on Whole Networks. *Journal of Management* 33: 479-516.

Ritter, T. and Gemünden, H.G. (2003). Network Competence: Its Impact on Innovation Success and Its Antecedents. *Journal of Business Research* 56: 745-755.

Ritter, T. et al. (2004). Managing in Complex Business Networks. *Industrial Marketing Management* 33: 175-183.

Rodan, S. and Galunic, C. (2004). More Than Network Structure: How Knowledge Heterogeneity Influences Managerial Performance and Innovativeness. *Strategic Management Journal* 25: 541-562.

Rowley, T. et al. (2000). Redundant Governance Structure: An Analysis of Structural and Relational Embeddedness in the Steel and Semiconductor Industries. *Strategic Management Journal* 21(3): 369-386.

Soda, G. et al. (2004) Network Memory: The Influence of Past and Current Networks on Performance. *Academy of management* 47 (6): 893-906.

Sporleder, T.L. and Moss, L.A.E. (2002). Knowledge Management in the Global Food System: Network Embeddedness and Social Capital. *American Journal of Agricultural Economics* 84 (5): 1345-1352.

Stank, T.P. et al. (2001). Performance Benefits of Supply Chain Logistical Integration. *Transportation Journal* 41 (2/3): 32-46.

Stuart, T.E. et al. (1999). Interorganizational Endorsements and the Performance of Entrepreneurial Ventures. *Administrative Science Quarterly* 44: 315-350.

Thorpe, R. et al. (2005). Using Knowledge within Small and Medium-Sized Firms: A Systematic Review of the Evidence. *International Journal of Management Reviews* 7 (4): 257-281.

Tsai, W. (2001). Knowledge Transfer in Intraorganizational Networks: Effects of Networks Position and Absorptive Capacity on Business Unit Innovation and Performance. *The Academy of Management Journal* 44 (5): 996-1004.

Van de Ven, A.H. (1976). On the Nature, Formation and Maintenance of Relations among Organizations. *Academy of Management Review* 1 (4): 24-36.

van der Vorst, J.G.A.J. (2006). Performance Measurement in Agrifood Supply Chain Networks: An Overview. In C.J.M. Ondersteijn et al. (eds.), *Quantifying the Agri-Food Supply Chain*, 15-26. Dordrecht: Springer.

Westerlund, M. et al. (2008). Sme Business Models in Global Competition: A Network Perspective. *International Journal Globalisation and Small Business* 2 (3): 342-358.

Wincent, J. et al. (2009). Network Board Continuity and Effectiveness of Open Innovation in Swedish Strategic Small-Firm Networks. *R&D Management* 39 (1): 55-67.

Zaheer, A. and Bell, G.G. (2005). Benefiting from Network Position: Firm Capabilities, Structural Holes, and Performance. *Strategic Management Journal* 26: 809-825.

PART II

CHAPTER ONE

THE LAS APPROACH: A SCHEME FOR A SUSTAINABLE LOCAL DEVELOPMENT OF LATIN AMERICAN RURAL AREAS?

DENIS REQUIER-DESJARDINS

Abstract

LAS have been viewed until now either as local production systems or as sets of links between food and places. Although these approaches are relevant, they do not always differentiate between close concepts such as Local production systems or "clusters". This chapter, drawing on a Latin American background, where LAS activation is viewed as a means to enhance local development processes, assesses in what measure the LAS approach can be redefined as a rural development approach, and therefore be a potential basis for local development policy.

If LAS is to be considered as the backbone of a rural development path, it must be cast in the current debate on diversification of rural activities and "new rurality". We believe that LAS can be the basis of a rural development policy, but may be fit only places with a specific "residential" relationship with the overall economic environment.

1. Introduction

The Localized Agri-food Systems (LAS) approach, which emerged in the late 1990s, is at first sight close to the cluster (or local productive system) approach, as it refers to geographical concentrations of specialized farms, food-processing units and distribution networks, private and public entities in a determined place. But some contributions (Muchnik et al. 2007; Requier-Desjardins 2010a) have tried to identify LAS specificity by underlining two characteristics:

- LAS relationship with agriculture and stock-breeding, which means a relationship with living resources; and
- food consumption status, characterized by proximity between the consumer and the product, hence the importance of qualification processes, in particular those based on geographical origin and reference to place.

These characteristics entail a specific link between LAS and place, less contingent than in the case of clusters, going further than the simple effect of geographical proximity or spillover effects. This leads some contributors to stress the food-place relationship rather than the LAS structure as a production system.

Nevertheless, at an analytical level, methodologies used for cluster analysis (e.g., network analysis to grasp their ability to innovate and diffuse knowledge, see Giuliani and Bell 2007) are relevant in the case of LAS and the overlapping of the two concepts can make the specificity of LAS approach somewhat fuzzy.

There is however another way to state LAS approach specificity: we can consider it as a scheme for a policy of sustainable local development in rural areas. This focus has emerged in the debate on the LAS approach carried on in Southern countries. Although it has wide application, the approach is particularly relevant for Latin America, where a significant number of LAS have been identified and where development bodies, such as IICA, set up a methodology of "LAS activation" for local development.[1] We shall therefore concentrate on this geographical area.

The purpose of this chapter is to elaborate on this dimension of the LAS approach, considering that it could lead to rural development public policy proposals. It will rest on the following assumptions:

- First, we cannot consider as an a priori statement, except in a very ideological way, that LAS represent a model of local sustainable development.
- Consequently we have to place LAS analytical approach into the framework of the debate on the meaning of development and on the forms rural development takes, particularly but not exclusively in Southern countries.

As LAS are about food-processing activities which are not agricultural and can be performed off-farm, we shall particularly stress the on-going debate on the diversification of activities in rural areas and the weight of rural non-farm (RNF) activities.

The chapter will thus focus on:

- a survey of the current debate on rural development and the trend towards "new rurality";
- recent theoretical approaches, able to deal with the current forms of rural development;
- the integration of LAS approach into this debate, in order to characterize it as a proposal for rural development.

2. Rural development and "new rurality": a brief assessment

There are two broad views of development:

- It can be characterized as a dynamic process of structural change of economies, engaged in a trend of growth of their per capita gross national product or income. This process is expected to provide welfare benefits.
- It can be characterized as a goal to be reached by that process and this goal is defined by a satisfactory level of welfare.

The first view is current in seminal contributions on development economics, but the second has currently become dominant for at least two reasons:

- First, the stress on development processes was in the past related to the design of national development strategies and models, "inward-looking", typical of pre-adjustment and the pre-globalization era. As national development paths and "models" are no longer relevant as research topics, structural change is today cast in a global framework, and like the focus on "sustainable development" is not specific to developing countries.
- Under-development today is defined by the lack of welfare for a significant part of the population, typical of Southern countries, in short, by the incidence of poverty. However the welfare approach, and poverty as lack of welfare, has been transformed by the stress put on "capabilities" and "good life" (Sen 1998), where freedom of choice and basic liberties become a component of development. The UN Millennium goals calling for the abatement of the different forms of poverty of capacities are an example.

However the reference to development as a process of structural change is not entirely ruled out by this new context. Even if the goal is no longer the construction of a coherent national economy in terms of inter-industrial relationship national strategies still target FDI which might boost growth and upgrade economic productive capacity, and their attractiveness is based on specificities in terms of resources, skills, organization, etc. They can be identified at national level, and have greater great relevance at local level. The localized nature of development processes is thus more conspicuous at a place level than at a national strategy level.

This is particularly obvious when taking into consideration literature on clusters and local productive systems, of which LAS literature can be considered a part. It is also obvious when taking into consideration the debate on rural development and the emergence of a "new rurality".

Since the 1990s, a broad range of contributions has stressed the weight of rural non-farm activities and employment in the income of rural households. According to FAO (1998), 50% of rural income comes from non-farm activities, the percentage being higher in Asia and Latin America (around 60%) but lower, though not minimal, in Sub-Saharan Africa (between 30% and 40%). More recent contributions confirm these figures. Davis et al. (2009), gathering data from RIGA data base over a subset of countries of Eastern Europe, Asia, Latin America and Africa find that in all countries but Vietnam and African countries, the share of off-farm income is above 50%. Moreover, there is a positive relationship between GDP per capita and the share of off-farm income, which tends to show that this characteristic can be seen as an indicator of development.

Some caveats are to be underlined in this respect: off-farm activities can be agricultural, when they concern wage-earning activities in capitalist farms or plantations; they can also be food-processing activities, as in the case of rural agro-industry, and thus keep a strong link with agriculture.

But these sources of income can also stem from activities no longer connected to agricultural production, such as handicrafts, tourism, industrial wage-earning, commerce and trade, some of these activities being linked to internal or international migration. The World Development Report 2008, the first to be dedicated to agricultural issues since the 1980s, stresses that the diversification of sources of income, such as wage-earning and migration, are ways out of poverty for small-holder, family-based agriculture.

There are conflicting analyses of this trend. Some (Davis et al. 2009) argue that agriculture remains the cornerstone of the set of rural activities, as diversification occurs mainly to cope with the uncertainties of agricultural production and income. They also stress the difference to be made at the meso-economic level of rural economy and at the micro-economic level of rural households: diversification of households at a meso- level, some of them quitting agriculture, is not the same thing that diversification of activities within households at a micro level. Others (Rigg 2006, drawing mainly on Asian case-studies) argue that given the weight of off-farm activities for rural households, agricultural development and rural development can be decoupled through the corresponding decoupling of indicators. Agricultural development may no longer be a way to mitigate poverty in rural areas. This can even lead to considering that migration is not an alternative to rural development, but development in its own right (Pritchett 2008).[2]

In any case this trend, which concerns Southern as well as Northern countries,[3] leads to the consideration of a "new rurality" which concerns much more than agricultural production. This is extremely clear when we consider the emergence of new forms of agricultural production, such as "network companies" (Clasadonte 2008). These are often linked to financial capital and fuelled by "knowledge economics": land is considered just one input among many and main actors are agricultural entrepreneurs rather than farmers (Hernandez 2008). This contributes to a further disconnection between agriculture and rural places, as agricultural production units are no longer defined by their access to land, which becomes a "generic" input.

This debate on new rurality is also leading to a redefinition of a rural place. Statistical definitions, based on density, vary across countries. But the issue is not only statistical: diversification of sources of income in rural households can refer to employment opportunities in nearby rural areas but also urban areas, and even, in the case of migration, in faraway places. Likewise, incomes paid by new forms of entrepreneurial agriculture, such as rents, can accrue to urban areas. Particular attention has to be directed towards small towns within rural areas, where rural population quitting agriculture might concentrate and develop new activities.[4]

The complexity of the issue of new forms of rural development calls for new theoretical approaches.

3. New forms of rural development: some theoretical approaches

As far as theoretical analysis is concerned, two theoretical approaches significantly change the view of the link between rural development and agricultural development:

- In terms of supply and activities, at the microeconomic level, the sustainable rural livelihoods approach stresses the risk aversion of rural households.
- In terms of supply and demand circle, at a more meso-economic level, the residential economies approach is based on "economic base theory".

The first approach refers to a vision of diversification of activities at the household level; the second refers to a rural economy level.[5]

3.1 The sustainable rural livelihoods approach

This sustainable rural livelihoods approach (SRL) can be considered as an offspring of the microeconomic producer-consumer modelling of agricultural households based on "New Household Economics" from the 1960s to the 1980s (Ellis 1993). These models were basically geared towards the discussion of microeconomic internal equilibrium of the household and its relation with general equilibrium and response to market incentives.

The SRL approach (Ellis and Freeman 2005) is more geared towards the appraising of long term sustainability of rural households and its link with overall sustainability of rural areas. Households are thus defined by their assets, or capital (productive, human, financial, natural and social) and their behaviour is viewed as the management of a portfolio of opportunities to mitigate risk and ensure the resilience of the household.

In this approach the various activities open to households are seen as alternatives; joint production effects are not emphasized. Therefore an approach in terms of productive systems could thus be useful to characterize the supply side of rural activities.

3.2 The residential economy approach

This approach can be considered as the transfer at place or territory level of the Keynesian approach of export multiplier.[6] It is grounded on the "economic base theory" (Andrews 1953) which states that the local economy can be divided into two very general sectors:

- A Basic Sector made up of local production units that are entirely dependent upon external factors. Manufacturing and local resource-oriented firms (like logging, mining or agriculture) are usually considered to be basic sector activities because their fortunes depend largely upon non-local factors as they "export" their production outside the local area.
- A Non-basic or "domestic" Sector composed of those firms that depend largely upon local business conditions. For example, local services are generally identified as non-basic because they depend almost entirely on local factors and demand.

That means that every place gets a certain level of income from the outside, its "exportation", which through the local demand this level of income entails, fuels a series of "domestic" activities. Consequently it is always possible to make a distinction between a territorial economic base and activities resulting from the multiplier effect of this base. It has to be stressed that these forward linkages are only demand linkages (and not for example spillover effects at the productive level).

According to Davezies (2008), who revised this approach to define a "residential" model of local development, four types of economic base can be identified:

- A productive base, local productive activities "exporting" outside the territory and distributing income to local people such as wages, interests or profits. It is the classical starting point of economic base theory.
- A residential base, local residents who enjoy incomes created and distributed in other territories. This can encompass for example the residence of retirees or the residence of professionals who work outside the territory, physically or through CIT. This can also include local tourism income.[7]
- A public base, the implementation of public administration and services and to the distribution of wages and subsidies it entails.
- A transfer base, linked to various forms of transfer made by

outside agents to local agents, including social security transfer or migrants remittances.

Obviously the main distinction is between the productive base and the three others, as these can overlap. What is the frontier between residence and part-time migration, for example? In the case of the productive base, incomes depend on external demand but are generated through local production; in the other cases incomes are generated outside but distributed locally.

This approach supports an economic model of new rurality. Rural development in many places is fuelled by the residential economy at large, meaning the multiplier effect of income transfer from outside. This concerns rural areas distant from urban economic centres and main transportation links, but boasting "amenities" likely to attract tourists or even residents such as retirees, or simply entitled to a significant amount of transfer money. It also covers rural areas close to urban areas with residents who commute.

The Residential Economy model has been mainly used for the characterization of the local development of e rural places in Europe (for example in France) which can register higher incomes per capita than urban areas concentrating industrial activities. So it allows for characterizing a new rurality where agriculture, namely the productive basis, is no longer dominant as the basis of the local economy.

As far as Southern countries are concerned, this move is only incipient, although some areas are already attracting tourism and urban residents. But the base of a residential economy in these countries is often incomes linked to migration, either incomes from part-time migration to urban areas or remittances of international migrants. As a matter of fact, migration and corresponding remittances show strong patterns of concentration in specific local areas (Requier-Desjardins 2009).

The Residential Economy approach is not only about rural areas: some cities can also attract residents and income generated elsewhere. In some cases incomes generated in agriculture can be distributed to urban residents, especially in small towns. This occurs in the "reverse tenancy" when small-holders rent their land to entrepreneurial agriculture as seems to be sometimes the case in Argentina for the soy sowing-pools.[8]

These two approaches offer a new vision or rural development. If development is the abatement of poverty, it is no longer based solely on agricultural income. If it is a specialization path, it is not necessarily geared towards agriculture or even agro-food.

Some crucial issues for the characterization of rural development paths can be drawn from these reflections.

The role of agriculture in the set of activities

As we have seen, some contributions to the debate (Rigg 2006) take a very radical view, claiming that the abatement of rural poverty and rural welfare in developing countries no longer depends on agricultural development, given the diversification of activities. Others claim that, though it may represent today less than 50% of the income of rural households, agriculture remains at the centre of rural livelihoods, thanks to its relationship with other sources of income, often viewed as means of tackling the variability of agricultural incomes, and which often have a productive synergy with agriculture.

The relationship with urban areas

Urban residents not only demand food products but enjoy a set of rural amenities either as tourists or residents. Place qualification can contribute to competitiveness of some agricultural products and attract residents and tourism. Products with some form of Geographical, Organic or Fair Trade Indication are mainly geared towards urban markets, in Southern countries too. But this mechanism rests on the capacity of income generation in urban areas geared towards these forms of consumption. This depends, particularly in Southern countries, on food consumption patterns linked to the existence of an urban upper-middle class. It depends on the capacity of rural areas to attract income generated in urban areas towards these forms of consumption, thus complete the circle of supply and demand locally.

The role of entrepreneurial agriculture in the shaping of income generation

In some countries, such as countries of the American South Cone, new forms of entrepreneurial agriculture emerge, from subcontracting agriculture to sowing pools and network companies, with intervention of financial actors and a decoupling of agricultural production from farm structure. Here, farmers can be reduced to small rent-seekers. Agriculture is akin to other sectors; knowledge, technological package and vertical integration being the main tenets of this new organization. Diversification of activities, including food-processing, concerns households whose link with agricultural production consists mainly in the perception of an income. Residence in rural areas is no longer a prerequisite to enjoy this kind of income. Diversification stemming from a multiplier effect of this income might occur in nearby urban areas.

All these issues must be taken into consideration when assessing LAS and the LAS approach for use in rural development proposals.

4. LAS and rural development

The doubts on the contribution of LAS to rural development stem primarily from the fact that they represent rural income generating activities. The main need is to fit them into a specific category. They probably appear in some cases within on-farm activities, especially in the case of rural agro-industry performed at farm level, but can also be integrated into off-farm manufacturing activities, especially in the case of the most developed LAS where there exist processing units managed by actors who are not primarily farmers.

However they are classified, their analysis can contribute to the design of adequate rural development policy Davis et al. (2009). We have identified three aspects of the link between LAS and the development path of rural areas:

- The microeconomic aspect of the portfolio management of activities by rural households (SRL approach for example). In this respect, rural agro-industry is a way to counter the low level of agricultural income, as in Latin America.
- The identification of an economic circuit, as used in macroeconomics, but here applied at a meso-economic level. This circuit stems from the distinction between an economic basis fuelling a demand, which can be residential, and domestic activities, linked to the base by a multiplier effect. Note that according to the "basket of goods" (Pecqueur 2001) approach, territorial qualification of products is based on demand and that this demand is largely fueled by external incomes.
- The intertwining of activities within local productive systems, and hence local agri-food systems, means that they need to be involved in any rural development policy. Rural development policy can only be based on the existence of synergies between activities. Some of these synergies stem from input-output links such as the link between rural agro-industry and agriculture. Other synergies stem from the sharing of the same territorial qualification characteristics such as the link between territorial qualification, agro-food and tourism, and others are potentially financial such as the link between remittances and the financing of LAS. But synergy can also from skills and competences of people which can be invested in the various activities.

In order to be considered as a part of a proposal for a rural development strategy, the LAS approach needs to encompass these three aspects. Their convergence should ensure the resilience of the system and the associated development path. Activities covered by LAS must include:

- An effective way to ensure sustainable livelihoods, meaning that they are part of the portfolio, not having been crowded out by other activities. Sustainable livelihoods can mitigate risks associated with the other activities in the portfolio.
- Development of productive and knowledge links, endogenous and exogenous with other activities present on the territory. Local systems and network enhancement.
- LAS must either be part of the territorial productive base or be effectively enhanced by the multiplier effect of an income distribution at the local level. This raises the issues of market access of LAS products and the link with tourism and people endowed with external money transfers, such as remittances.

At the micro-economic level, rural agro-industry status deals with the first constraint, in as much as it can provide farmers with an opportunity to develop a food-processing activity within the household. But as we have seen, Davis et al. (2009) have stressed the difference between diversification of activities in a rural area at the meso-level and diversification of activities at the micro level of households: only the second type of diversification can help households to mitigate poverty and vulnerability through portfolio management of this set of activities.

In the early days of rural agro-industry (Boucher and Riveros 1989) diversification was seen mainly as a way for smallholders to gain more value added by climbing downstream along the processing chain of agricultural products. But, as LAS developed, the picture changed. In many cases LAS production units are no longer embedded in farms and even where they are, food-processing units might well represent the cornerstone of livelihood structure, and agriculture be relegated to a second level. This evolution entails heterogeneity of actors involved in LAS, some of them being in a position to control the system and gain most of its value added. LAS networks present nodes where key information about market opportunities, technology, new products, etc. transit and which in particular manage relationships with external actors. LAS literature provides a number of examples:

- In Colombia (Requier-Desjardins and Rodriguez-Borray 2004), "trapiches" (sugar cane processing units) are the main actors of "panela" "Hoya del Rio Suarez" LAS. If "trapicheros" are also sugar cane cultivators, they process sugar cane coming from various smallholders, and are more geared towards this entrepreneurial activity than towards sugar cane cultivation as such.
- In Cajamarca (Boucher 2004) urban cheese-mongers of Cajamarca are at the centre of the system as they have been able to design a collective territorial trade mark. Most are urban dwellers with a higher level of education, and are sometimes former civil servants hit by the "Fujishock" and searching for other livelihoods. They control networks of milk suppliers, and to guarantee the quality of the milk they process and control the market in Cajamarca and supermarket access. These hierarchical networks entail a process of selection among milk producers as well as an exclusion of small scale cheese street-sellers.
- In Turrialba (Blanco and Riveros 2004), there are two groups of cheese processing units, small ones generally integrated into farms and independent "industrial" SMEs. These two groups have very different strategies regarding the qualification of Turrialba cheese: the small ones are trying to develop a "Turrialba cheese route" which would allow them to widen the scope of their livelihoods. The SMEs are more interested in securing external markets by maintaining relationships with urban supermarkets.

As a result, food-processing activities might well not represent a diversification of activities for small-holders at the microeconomic level. For example milk processing, in order to enforce standards of hygiene requested by a cheese producer, hardly represents diversification for a cattle-breeder. Market outlet diversification for smallholders (e.g., milk sales or cheese) is nevertheless part of the process of diversification. It is far from sure that LAS entails an abatement of poverty for excluded groups (Requier-Desjardins 2010a). Successful LAS might represent an opportunity for smallholders to integrate into a network of providers of local food-processing activities; under the condition they enforce standards regarding quality of raw material (the "*queso mantecoso mejorado*" in Cajamarca for example). Whether these opportunities are more valuable than "classic" contract agriculture managed by large firms is an open question.

The structure of the LAS production system and the way LAS

activities are set into the local economic circuit can be examined at meso-economic level.

As far as the production system is concerned, the process of "activation" of specific assets involves collective action, and this constitutes the core of "LAS approach" as a methodology for local development. Usually the process involves an external actor, a development promoter such as a public organization, a university or technical centre, etc., and a local committee which is set up through participatory methods such as SWOT analysis, participatory mapping, etc. But collective action is underpinned by important networks (productive, social or knowledge networks) and their capacity to enhance collective action, particularly through innovation, as stressed by Giuliani and Bell (2007). Networks of actors can spread the information about new technology, organization or market opportunities. However the LAS approach stresses the different positions of stakeholders in networks depending on their degree of centrality and their possible role as "gatekeepers". The structure of networks impacts the capacity for collective action, and in some cases inhibit it.

LAS cannot of course be considered as systems devoted to one type of activity. There exist input-output relationships, for example the processing of milk into cheese or of sugar cane into *panela*, and there is also the intertwining of activities sharing the same characteristic of quality, according to the "basket of goods" model. Many LAS in Latin America in fact show a link between food processing of quality products and tourism: this has been noted for Cajamarca (Peru), Turrialba (Costa-Rica) or Salinas (Ecuador). Tourism is a way to ensure a market for LAS products because it underlines their geographical specificity, and tourists are likely to continue consuming theses products at home. Furthermore there is a significant development of amenity-based rural tourism in many Latin American countries, which in some cases has rehabilitated farm buildings (*Haciendas* in Mexico, *fincas* in the *Eje cafetero* in Colombia for example). This kind of link between activities sharing the same territorial quality characteristics refers to the structure of the local productive system, but in fact depends on demand, given that quality needs an agreement between the consumer and the producer. Demand will make up the "basket" by selecting products which it perceives to boast the territorial quality characteristic (Pecqueur 2001). The basket of goods is created and made up by demand, mostly mostly but not entirely based on tourism. Some products of the area are recognized as enjoying this territorial quality characteristic, others are not. Demand might be external but it has to do at least primarily with an internal perception of quality at the territorial level. The role of demand rests on the needs of a consumer persuaded of the relevance of a local quality characteristic, which is facilitated by geographical proximity. In this case, demand has a mainly qualitative role of selection, but its quantitative aspect also needs to be addressed. Assessment of the local dynamics of supply and demand requires the assessment of demand and the mechanics of income generation.

The role of demand and the way it is fed by the generation of income has not been until now a central concern in LAS literature. However, LAS are able to generate incomes, and meet market demand expressed by income earners.

As far as generation of income is concerned, as the LAS approach stems at least partly from the consideration of rural agro-industry as a way to of raising the part of value added at farm level, it might be predicted to improve the income of marginalized rural areas and therefore implicitly the level of economic activity stemming from the use of this income. But the shift from rural agro-industry to Localized Agri-food Systems entails the recognition of the heterogeneity of actors and thus the discrepancies in their access to incomes flowing from these systems. Part of this income might well accrue to external actors. The consolidation of a true LAS system might well represent a greater difficulty for smallholders, and particularly the poorest, in accessing "Rural Non-Farm" activities.[9]

Moreover, LAS can be part of the place-specific productive base, with a local multiplier effect of the incomes it generates. The capacity to export outside the territory is mainly driven by its capacity of entry into distribution networks down the commodity chain, such as supermarkets, which usually require compliance with quality norms. Moreover if their specificity needs to be recognized at the national or international level a reliable quality marker (GI, collective trade mark, etc.) might be necessary.[10] Otherwise, given the link with territorial quality and the role of proximity with the consumer, market access can depend on a residential base. In the case of tourism or residents attracted by amenities, these activities are "domestic" in the framework of a residential/presence economy, meaning that a multiplier effect of the economic base is at work to ensure the level of demand. The issue of the characterization of territorial economic base however remains open.

In the case of tourism, incomes are mainly of urban origin, probably mostly earned by upper-medium income urban residents. But the economic base can also derive from transfers, which might be considerable given development of conditional cash transfers in some Latin American countries.[11] Given their importance for rural areas in many Latin American countries we would like to focus on migrants' remittances.

Migration, which is part of the microeconomic portfolio of opportunities, can participate in the possible financing of LAS activities, and migrant networks can also be effective in transferring skills. But remittances are mainly the source of an external residential income. There is currently a wide debate about the impact of remittances on development. It particularly concerns the local level of rural territories (Requier-Desjardins 2010b), given that migration, as remarked above, shows a largely territorial structure, with a concentration of flows at the origin as well as the destination. Investment generally represents a minor although not negligible part of the use of remittances. But in rural areas, it seems that the impact on investment in agriculture of food-processing activities is largely predetermined by competitiveness and abilities in agricultural and food production. The evidence regarding links between migrant remittances and LAS development is sketchy (e.g. Guzman Hernandez et al. 2009) but it seems that some exportation flows of LAS products (*panela* from Colombia, cheese from Mexico) are very much the consequence of the existence of a diaspora market in migration countries (United States for Latin America). Likewise, according to Ribaï (2010), drawing on an Ecuadorian case-study, it seems that migration, which can crowd out subsistence agriculture as a livelihood in marginalized areas, can be strongly associated with the development of commercial agriculture and food-processing by rural households in areas close to an urban market. This means that, when there is a strong productive base supplying close urban areas, the residential base, or remittances, is able to support this productive base through the development of a household-located financial service. Either by investment or consumption, remittances are likely to spur demand through a multiplier effect. Their level of concentration in some territories can make up a dominant residential base for these territories. LAS in this type of context should basically be considered as a "domestic" activity.

So a development path based on LAS dynamics needs to consider the production structure and the demand side, at the level of the local supply and demand economic circuit, rather than simply the characteristics of quality. Either LAS can be part of the productive base, as it "exports" outside the territory, or LAS is dependent of a residential/presence demand fueled by external incomes. This aspect is affected by the need for recognition of geographical origin quality characteristics.

5. Conclusion

We would like to emphasize three conclusions which can be drawn from the preceding analysis.

First it appears that LAS approach can represent a basis for a rural development policy mainly in the framework of a "new rurality trend", marked by diversification, relative disconnection between agriculture and rural households and new "residential" functions of rural areas. Its main advantage is that it entails a form of diversification which makes explicit the link between agriculture and rural non-farm activities, thus reinforcing the connection between rural development and agricultural development.

As it deals more with meso-diversification of rural places, the approach may represent a viable local development path, although it may not necessarily be a way of enforcing resilience and sustainability of rural households at a micro level. It cannot be considered as the only policy of abatement of rural poverty in concerned marginalized rural areas, or even as the most appropriate.

Moreover, given that it entails a residential link with urban areas at various levels, it only fits the situation of rural places that are characterized by the dominance of smallholders and a network of relationships with urban places. LAS development appears to depend on economic relationships at the macro-meso level, and cannot thus be seen as a general alternative proposal to other forms of rural or agricultural development.

Notes

1. For LAS methodology see IICA, CIRAD and REDSIAL in Mexico (Boucher, Reyes Gonzalez 2011; Chavez Rosas 2010).
2. Rigg (2006) states that in many Asian rural areas, people do not anticipate their income from agriculture; Pritchett (2009) considers that migration is a component of development, income per national being higher than income per resident.
3. In France only 15% of rural active population works in agriculture. Likewise in Mexico, as stated by Carton de Grammont (2009), 50% of rural active population does not work in agriculture.
4. For example, in the Argentine Pampas, where soy "sowing pools" have developed, there has been a recent process of concentration of rural population in small towns (Guardia and Tomarolli 2009).
5. Davis et al. (2009) note that the assessment of the diversification of activities might not be the same at both levels.

6. Note that Lucas (1988) stresses that the mechanisms of economic development are not essentially different at a national level or at a more reduced geographical scale.
7. As tourists are not residents, the term "presence" economy is used in this paper.
8. According to Elverdin et al. (2008) some of these "small landlords" might use their rent to develop food-processing units of quality products in these small towns.
9. Davis et al. (2009) express a concern about the unequal access to RNF activities for rural households, which can deepen poverty and inequality.
10. Cajamarca cheese LAS (Boucher 2004) is a good example.
11. PROCAMPO and Oportunidades in Mexico or Bolsa Familia in Brazil, for example.

References

Andrews, R.B. (1953). Mechanics of the Urban Economic Base: Historical Development of the Base Concept. *Land Economics* 29: 161-167.

Blanco, M. and Riveros, H. (2004). *La ruta turística del queso Turrialba.* Ponencia presentada a la conferencia ARTE (Agro-Industria Rural y Territorios), UAEMEX, Toluca, diciembre de 2004.

Boucher, F. (2004). *Enjeux et difficultés d'une stratégie d'activation des concentrations d'agro-industries rurales le cas des fromageries rurales de Cajamarca (Pérou).* Thèse pour le doctorat de Sciences Economiques, Université de Versailles Saint Quentin en Yvelines, décembre 2004.

Boucher, F. and Reyes Gonzalez, J.A. (2011). *Guía metodológica para la activación de sistemas agroalimentarios localizados.* Mexico City: IICA/CIRAD/REDSIAL.

Chavez Rosas, E.H. (2010). *Diagnostico de las agroindustrias rurales de la Selva Lacandona, Chiapas, México.* Mexico City: REDSIAL, cuaderno de Trabajo n. 2.

Clasadonte, L. (2008). *Network companies, another way on thinking agriculture. A supply chain management vision in South America.* Master thesis, Wageningen University.

Carton de Grammont, H. (2009). México: boom agrícola o persistencia de la pobreza rural. In *Boom agrícola o persistencia de la pobreza rural, estudio de ocho casos.* Santiago de Chile: FAO.

Davezies, L. (2009). L'économie locale "résidentielle". *Géographie, Economie Sociétés* 11 (1): 47-53.

—. (2008). *La France et ses territoires, la circulation invisible des richesses.* Paris: Seuil.

Davis, B. et al. (2009). A cross-country comparison of rural income-generating activities. *World Development*, 38 (1): 48-63.

Ellis, F. (1993). *Peasant Economics*. Oxford: Oxford University Press.

Ellis, F. and Freeman, H.A. (2005). *Rural livelihoods and poverty reduction policies*. London: Routledge.

Elverdin, J., Maggio, A. and Muchnick, J. (2008). Procesos de localización / deslocalización de las actividades productivas: expansión sojera y retracción ganadera en Argentina: estrategias de los productores. Paper presented at the 4th international congress SYAL ALFATER, Mar del Plata, October 2008.

FAO (1998). *The state of food and agriculture: rural non-farm income in developing countries*. Roma: Food and Agriculture Organization.

Giuliani, E. and Bell, M. (2007). Catching up in the global wine industry: innovation systems, cluster knowledge networks and firm-level capabilities in Italy and Chile. *International Journal of Technology and Globalisation* 3 (2/3): 197-223.

Guardia, L. and Tomarolli, L. (2009). Boom agrícola o persistencia de la pobreza rura en Argentinal. In Graziano Da Silva, J., Sergio Gómez, E. and Rodrigo Castañeda S. (eds.), *Boom agrícola o persistencia de la pobreza rural, estudio de ocho casos*. Santiago de Chile: FAO.

Guzman Hernandez, E. et al. (2009). Las redes sociales para la producción de queso añejo en la comunidad transnacional de Laguna Grande, municipio de Monte Escobe, Zacatecas México. In Espinoso Ortega, A., Cervantes Escoto, F. and Espinosa Ayala, E. (eds.), *Balance y perspectivas del campo mexicano*. Vol. II, *Globalización y respuestas locales de la agroindustria*. México City: Juan Pablos Editor.

Hernandez, V.A. (2007). El fenómeno económico y cultural del boom de la soja y el empresariado innovador. *Desarrollo Económico* 47 (187): 331-365.

Lucas, R. (1988). On the mechanics of economic development. *Journal of monetary economics* 22 (1): 3-42.

Muchnik et al. (2007). Introduction: Les Systèmes agroalimentaires localisés (SYAL), dossier «Systèmes agroalimentaires localisés», *Economies et Sociétés (série agroalimentaire)* 29: 1465-1484.

Pecqueur, B. (2001). Qualité et développement territorial: l'hypothèse du panier de biens et de services territorializes. *Economie rurale* 261: 37-53.

Pritchett, L. (2008). *Income per natural: measuring development as if people mattered more than places.* Working paper 143. Harvard: Center for Global Development.

Requier-Desjardins, D. (2009). La contribution des transferts de migrants au développement en Amérique Latine: une approche par l'économie résidentielle. Communication au 46ème colloque de l'ASRDLF. *Entre projets locaux de développement et globalisation de l'économie, quels équilibres pour les espaces régionaux*. Clermont-Ferrand, Juillet 2009.

—. (2010a) L'évolution du débat sur les SYAL: le regard d'un économiste. *Revue d'Economie Régionale et Urbaine* 4.

—. (2010b). *International migration from Southern countries rural areas: which impact on agricultural and rural sustainability?* Paper presented at ISDA Conference, Montpellier.

Requier-Desjardins, D., Rodriguez-Borray, G. (2004). Environmental impact of panela food-processing industry: sustainable agriculture and local agri-food production systems. *International Journal of Sustainable Development* 7 (3): 237-256.

Ribaï, N. (2010). *Del huerto a la ciudad: agricultura familiar y aprovisionamento urbano en la sierra ecuadoriana*. Paper presented at the CEISAL conference, Toulouse, July 2010.

Rigg, J. (2006). Land, farming, livelihoods and poverty: rethinking the links in the rural south. *World Development* 34 (1): 180-202.

Sen, A. (1999). *Development as Freedom*. Oxford: Oxford University Press.

CHAPTER TWO

ENVIRONMENTAL AND TERRITORIAL PROBLEMS OF THE SPANISH OLIVE OIL SECTOR AND PRIORITIES FOR RESEARCH AND INNOVATION: A DELPHI ANALYSIS

JAVIER SANZ-CAÑADA, DANIEL COQ-HUELVA, FLORENCIO SÁNCHEZ-ESCOBAR AND ISABEL HERVÁS-FERNÁNDEZ

Abstract

Spanish olive oil LAFS host a wide environmental diversity and a great variety of olive grove management systems. The types of collective action and territorial governance also differ significantly from one LAFS to another.

The main objectives of this chapter are: i) to analyse and evaluate the main environmental and territorial problems affecting the olive oil LAFS; ii) to assess research and innovation priorities that the Spanish R&D&I system must enhance; iii) to examine the relationship between the problems and the research programmes. The methodology is supported by a cross-interdisciplinary approach, based on Social, Agricultural and Environmental Sciences. A Delphi analysis is applied to a panel of experts in order to evaluate the problems and the programmes. The quantitative scores are supplemented by qualitative information obtained from a round table of experts and from the comments made through the Delphi survey.

As a result, seven R&D&I programmes, grouped under two big categories, are established: i) agro-environmental externalities corrective actions; ii) actions that address the olive oil LAFS and its multifunctionality. These programmes are harmonised with the main problem categories. Soil erosion and degradation appears as the most important problem, whilst changes in the olive grove geography also emerge as a key territorial issue.

1. Introduction and theoretical framework

Spanish olive oil *Local Agro-food Systems (LAFS)* (Économies et Sociétés 2007; Muchnik, Sanz-Cañada and Torres-Salcido 2008; Muchnik and Sainte-Marie 2009; Sanz-Cañada and Muchnik 2011) are characterised by their high levels of environmental and territorial diversity. In Spain, olive groves occupy more than 2.5 million hectares of land. These groves have a wide range of agro-ecological features, landscapes, productivity levels and management systems. Experts have characterised three main groups of olive farms in Spain: intensive (500,000 hectares), average-yield (1.2 million hectares), and low-yield farms (800,000 hectares) (Sanz-Cañada et al. 2008). In the first group, the increase of production, which is concentrated particularly on new areas of irrigated intensive or super-intensive olive grounds, exerts a pressure towards the reduction of bulk olive oil prices in international markets. In the second group, average-yield olive groves, un-irrigated and traditional orchards, must be restructured to remain competitive; this is particularly important in regards to creating strategies that could reduce the cost of harvesting. In the third group, low-yield olive groves in sloping land, also un-irrigated and traditional orchards, are generally considered as economically marginal, although they often retain relatively high environmental, landscape and cultural values in the mountain areas.

Moreover, the territorial organisation of Spanish olive oil LAFS often varies widely. In various areas in which olive oil production is the predominant source of income, it is possible to observe a high spatial concentration of entrepreneurial and institutional networks of olive farms, olive oil milling companies (most of them cooperatives), marketing firms, suppliers of inputs, concentric diversification companies (soaps, woodcrafts...), etc., in addition to a significant number of other institutions, including: associations for employing agricultural engineers for the integrated production, organic producer networks and associations, regulatory and inspection boards that monitor protected designations of origin, local action groups, etc. As a result of the existing high economic, ecological and social diversity, the types of collective action and territorial governance differ significantly from one LAFS to another. So LAFS' effects on local olive chains and territorial development vary widely.

From another point of view, new public policy approaches, as can be seen in the 2013 Common Agricultural Policy reform, will strengthen the monetary incomes associated with the provision of public goods. Not only do olive LAFS produce commercial goods and add value to local chains, but they are also characterised by their high capacities for multifunctionality.

Firstly, agro-environmental policies aim to support olive production systems that reduce negative agro-environmental externalities, such as erosion or diffuse pollution, and also to promote the generation of positive agro-environmental externalities, such as the carbon sink effect or the maintaining of biodiversity. Secondly, although the objective of commercial enhancement strategies is focused on obtaining differentiation-related incomes from high quality oil, these LAFS are also the subject of farming and rural development policies, as they generate positive externalities, such as the conservation of local olive heritage and the creation of a local network for the dissemination of innovation and knowledge. Finally, the economic valorisation of wastes generated in the olive oil chain is not only increasingly considered as a source of additional income, but also a strategy for optimising externalities (renewable energy, restoration of degraded soils, etc.). They are thus also targets of public policies.

As a result, the management of spatially different ecological, social and economic features and the implementation of multifunctional policies, require a new territorialised focus on research, innovation and transfer programmes. Consequently, to tackle this issue, it is first necessary to identify the main environmental and territorial problems concerning the Spanish olive oil LAFS. The analysis and assessment of these problems and the R&D&I priorities are the main objectives of this study. This approach is based on Social, Agricultural and Environmental Sciences. This study is part of a wider research project on research and innovation priorities in the Spanish olive oil sector (Sanz-Cañada et al. 2008), that uses a combination of complementary social research techniques applied to panels of experts.

This multidisciplinary study is directed at two main areas of knowledge in regards to sustainability of the olive oil chain and the territory: i) the *environmental externalities of the olive oil chain*, and ii) *multifunctionality, territorial governance and rural development.*

The first topic entails multidisciplinary research and innovation in various interrelated elements that are subject to high levels of territorial variability: fighting against soil erosion and degradation; improvement of water use in olive grove irrigation; reduction of soil and aquifer pollution by means of environmentally friendly farming systems; preservation of biodiversity and olive variety heritage; sustainability of oil milling activities; and waste research, especially in regard to olive pomace and pruning (García-Azcárate, Martin and Orlandi 2010; Gómez-Calero 2009; Guzmán-Álvarez 2004).

The second topic covers a wide range of aspects relating to territory, multifunctionality, sustainability, natural and cultural heritage, governance and rural development (Arriaza and Nekhay 2010; Sanz-Cañada 2009; Sanz-Cañada and Macías 2005). It was during the final years of the twentieth century that agricultural multifunctionality became progressively more frequent in public policy debates. Normative approaches look at problems of multifunctionality from a multidisciplinary perspective and are aimed at agricultural and rural development policies. They are based on the assertion that efficiency must not be the sole parameter for agricultural policies and that the emphasis should be placed on a range of environmental, economic and social criteria (Cairol et al. 2009; Mann and Wüstemann 2008; Renting et al. 2009; Stobelaar et al. 2009; Zander et al. 2007). Multifunctionality approaches are especially recommended to provide solutions to issues related with marginal olive groves. The wide distribution of marginal olive groves, their risk of abandonment and the imminent changes in the olive grove geography are indicators of the large territorial scale of the phenomenon.

In Section II, the methodology for this study will be presented. In Sections III and IV, the results of the Delphi analysis applied to expert panels will be discussed, including: i) the environmental and territorial problems of Spanish olive oil LAFS (Section III); *ii*) the priorities in research programs and lines, as well as the relationship between research programmes and environmental and territorial problems (Section IV). The conclusions will be presented in Section V.

2. Methodology

The project that frames this study addresses research and innovation priorities in the Spanish olive oil sector and has various objectives: i) identify the main problems of Spain olive oil LAFS; ii) identify and define lines of research and innovation that would help to strengthen the Spanish R&D&I system for olive oil and olive groves; iii) establish priorities for the above; iv) provide a series of strategic recommendations for implementation of R&D&I policies in the Spanish olive oil sector.

The methodology of the whole project is based on a sequence of three consecutive phases (figure 2.1). The first consists of identifying and defining 86 lines of research and innovation based on long semi-structured interviews with experts. The second involved three expert panels that used a workshop strategy (a short Delphi survey, a discussion group and a strategic participation workshop based on the Metaplan methodology) to achieve a consensual vision of the problem, a selection and categorisation of the research lines and a structured series of strategic recommendations for improving the R&D&I system (Oakley 1991). These first phases were focused on areas of longitudinal-disciplinary knowledge: i) olive farming and olive wastes; ii) milling technologies and new products; iii) olive oil Economic and Social Sciences.

The first two phases confirmed as follows: i) the need to apply more sophisticated discriminating tools to the consensuses reached; ii) the need to supplement the disciplinary approach with a cross-interdisciplinary analysis, based on interrelating technical knowledge (agricultural, agro-industrial, ecological...) and Social Sciences knowledge. In response to the above, the objective of the third phase was to establish a second, more in-depth, consensus among a great number of experts, in order to define priorities in regards to the problems and research lines established in the previous phases. Two cross-interdisciplinary *Delphi analyses* to two thematic areas, involving 70 R&D&I lines, were completed: i) the value chain, quality, food security and consumption, that deals with a vertical food chain focus; ii) sustainability, territory and rural development, that deals with a horizontal and territorial approach. The expert selection, which involved a total of 130 experts, was made seeking to obtain a balanced mix among the different research profiles (scientific, business, public administration, etc.).

This chapter discusses some of the results obtained from the second *Delphi analysis on sustainability, territory and rural development* (outlined in Figure 2-1). A Classical Delphi model design was used. Its objective was to establish a consensus and also to predict future tendencies and events. The study was based on the assumption that the selected experts had an extensive professional experience and great information in order to establish a series of shared priorities. For this reason, it was not deemed necessary to introduce weighting coefficients in the measures of central tendency.

Answers to the Delphi survey questions were based on the Likert scale: 5 was the highest possible mark ("very high importance" or "very high certainty") whilst 1 was the lowest possible mark ("very little importance" or "very little certainty"). Additionally qualitative information was obtained, in part, from comments made by the experts in the Delphi survey stage and, also, during a round table, which took the format of an expert discussion group.

To define the level of consensus, the inter-quartile range (IQR) was used. That is to say, for a given question a "high consensus" was achieved when the IRQ $\leq$ 1. For many questions (69.33% of the total) a high consensus was achieved in the first round of the survey. The fact that lines of research and innovation would have been endorsed by a progressive selection process in the first and second phases of the project facilitated the consensus process, which was obtained in the second round for all items. The experts' response rate during the two rounds was 58.3% and the aim of attaining 35 valid surveys was fulfilled following the two rounds.

Figure 2-1 R&D&I priorities, opportunities and problems in the Spanish olive oil sector: methodological framework

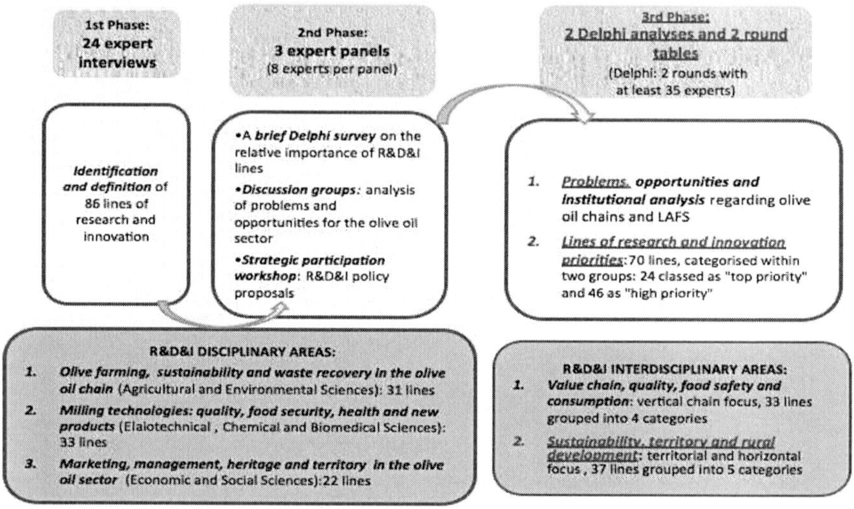

3. Environmental and territorial problems of Spanish olive oil LAFS

The experts were provided with a list of fourteen main problems. Their definition and median are shown in Table 2-1. The arithmetic mean of the problems is presented, in decreasing order, in Figure 2-2. Depending on the scores and qualitative information supplied by the experts, these problems were divided into four sub-groups that are examined in the next pages.

Table 2-1 Main environmental and territorial problems of olive oil LAFS
Median score of "importance of the problem": Me IP

	Me IP	Abbreviation
High level of soil erosion and degradation, especially in the case of olive groves on sloping land	5	EROSION
Excessive use of synthetic phytosanitary products (herbicides, pesticides, fungicides, etc.)	4	FITOS
Widespread verticillium wilt amongst irrigated olive orchards	4	VERTICIL
Organic weakening of soil and scarce availability of organic fertilisers	3	FERTIL
Appearance of new environmental problems as a result of super-intensive olive production (pests and diseases, use of agrochemicals, plant strength, loss of biodiversity...)	3.5	SUPERINT
Optimisation of water use in irrigated farms	4	RIEGO
Limitations on water resource availability for olive production due to the increased irrigation	4	HIDR
Loss of biodiversity in olive grove agro-systems	4	BIODIV
Regression of autochthonous olive varieties	3	TIPIC
Geographical olive grove changes as a result of increased intensive and super-intensive cultivation	4	GEOGR
Cost reduction requirements for extensive olive production	4	COSTES
Need to restructure medium and low-yield olive groves	4	REESTR
Lack of innovation in regards to machinery adapted to traditional olive groves on sloping land	4	MAQUIN
Scarce efforts to add value to the environmental, social and cultural roles of the olive oil LAFS	4	MULTIF

1= very little importance; 2 = little importance; 3 = average importance; 4 = high importance; 5= very high importance.

Figure 2-2 Main environmental and territorial problems of olive oil LAFS
Arithmetic mean score of "importance of the problem"

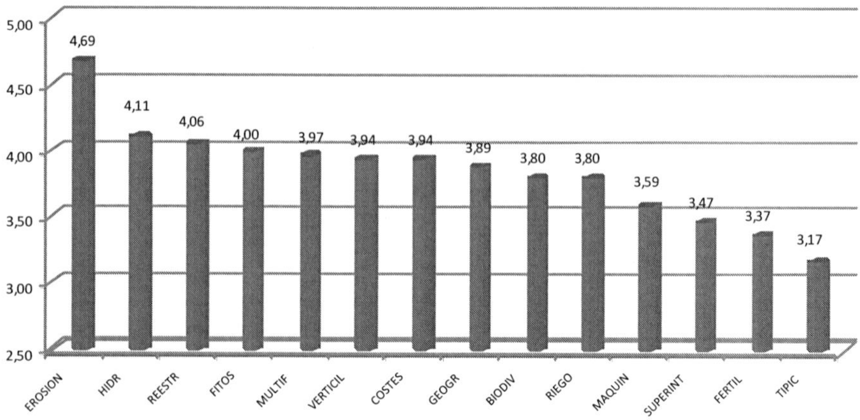

1= very little importance; 2 = little importance; 3 = average importance; 4 = high importance; 5= very high importance.

3.1 Environmental problems regarding soil erosion and degradation, diffuse pollution and biodiversity

The *high level of soil erosion and degradation, especially in the case of olive groves on sloping land* (EROSION) represents an environmental and territorial problem to which the experts awarded the highest average score of all the fourteen problems (μ=4.69). It is also the only problem attaining a median of 5, which means that it is a "very important problem". The experts highlighted that inadequate soil management is a significant factor in olive grove soil degradation. This is in part attributable to the general systematic use of herbicides to maintain bare soil, which contributes to the destruction of adventitious plant coverage, but also due to the failure to consider natural organic fertilisation processes (Calatrava et al. 2007; Francia et al. 2006; Graaff et al. 2010).

The *excessive use of synthetic phyto-sanitary products (herbicides, pesticides, fungicides, etc.)* is a closely related issue, identified by the experts as of "high importance" (FITOS; μ=4 & Me=4). Amongst the main environmental externalities that are caused by these practices, nitrate contamination of ground and surface water and soil fertility reduction are of particular concern; these effects are evident in regards to water retention and increased erosion. Furthermore, the use of herbicides can cause negative externalities not just in the surrounding land, but also in oil milling, as this entails additional olive washing and water treatment costs.

The *loss of biodiversity in olive grove agro-systems* (BIODIV, μ=3.80) is another important issue according to the experts (Me=4). It is also recognised as an aspect that is closely related to olive grove management systems. Therefore, certain common practices, such as the use of herbicides or copper salts, and the maintenance of bare soils without any supplying of organic matter, have a serious impact on biodiversity of agro-systems. Improving biodiversity is especially important when addressing management of olive groves that are likely to be abandoned.

Finally, two problems were designated to be of "average importance" (Me=3). The first of these was the *organic weakening of soil and scarce availability of organic fertilisers* (FERTIL, μ=3.37). Although the experts recognised that organic fertilisers external to the farm are scarce, they considered that the low availability of these resources does not represent a major problem because a more holistic solution is needed for preventing soil degradation, including fertility problems; this focus seemed to contribute to assigning a median of 3. The experts proposed enriching the soil with pomace and pruning waste and the use of vegetable covers in the crop. Finally, although we believe that the *regression of autochthonous olive varieties* (TIPIC, μ=3.17) restricts current and future possibilities for local differentiation of olive oils, the experts have only attributed a median of 3 to this problem.

3.2 Environmental problems regarding water resources for olive groves

The *limitations on water resource availability for olive production due to the increased irrigation* (HIDR, μ=4.11) and the *optimisation of water use in irrigated farms* (RIEGO, μ=3.8) were both considered as being of "high importance" by the experts (Me=4). The expansion of intensive and super-intensive olive farming has resulted in a significant increase in water used, despite generalised improvements in irrigation efficiency. The panel was of the opinion that in many areas in which the development of irrigated intensive farming system is particularly strong, demand will soon exceed the water supply availability. In 2008, irrigated olive groves occupied 418,170 hectares in Spain; this represented 17% of the olive groves surface (MARM 2009).

The panel also recognised the need to optimise irrigation use, although the experts were initially divided in this topic, as was pointed out in the qualitative information provided. Some experts believed that water use in olive groves irrigation is already sufficiently developed and that current localised irrigation systems are efficient. On the other hand, other experts stated that, as irrigation water is provided as part of a government concession which is not proportional to the volume consumed, farmers externalise a significant part of the cost attributable to its use. It appears that knowledge of deficit irrigation techniques and rational use of water amongst farmers is not sufficiently developed.

3.3 Problems regarding the need for restructuring and changes in olive grove geography

Many of the experts commented that most of the main environmental and territorial issues of olive oil LAFS have been catalysed by common agricultural policies that historically favoured, until the CMO reforms in 2003-04, the intensification of production, with no regard to the environmental consequences. On the other hand, although EU agro-environmental programmes were established in 1992, financial funds directed to support traditional olive grove systems that could foster positive externalities have been much lower when compared to market support policies. The panel also identified the lack of an integrated and public system of agricultural advisory services that could connect farmers with the R&D&I system and counteract the strong influence that agricultural input companies have traditionally had on farmers.

Five problems have been considered as part of this section, all of which have a median of 4 (high importance). The difference registered between each of the four problems is relatively low, with all attaining a score between 3.89 and 4.

The first two address the restructuring process of extensive olive groves: *need to restructure medium and low-yield olive groves* (REESTR, μ=4.06) and *cost reduction requirements for extensive olive production* (COSTES, μ=3.94). The panel stated that the reconversion of average-yield olive groves can be achieved by one of two means. The first option is to convert existing orchards to intensive olive groves, but this would be difficult to implement on a wide scale due to the lack of new water resources in many areas. The second option, which is well suited to groves on a relatively flat surface, is to select strategies of reduced harvesting costs, which represent the main cost incurred by olive farms. According to AEMO (2010), harvesting costs represent between 35% and 41% of the total costs of the farms, depending on the type of olive grove.

Problems associated with *geographical olive grove changes as a result of increased intensive and super-intensive cultivation* (GEOGR, μ=3.89), are closely related to the two problems mentioned above. This fact is causing a downward trend in the international price of bulk olive oil. When adding the effects of the strengthening of the commercial margins in the olive oil chain promoted by the big retailing companies, the opportunities for traditional mountain low-yield olive groves to make a profit are limited. The experts stated that unless public policies are implemented, a significant part of marginal Spanish olive groves may be abandoned. In areas of traditional olive orchards on sloping land, there are often no agricultural alternatives to olive farming. Therefore, if these groves fell into disuse, the consequences could be irreparable, from both an environmental perspective (biodiversity, increased risk of forest fires, etc.) and a socio-economic and territorial viewpoint (depopulation of rural areas, disappearance of economic activities, etc.) (Fleskens and Graaff 2008; Guzmán-Alvarez and Navarro 2008; Nekhay, Arriaza and Guzmán-Alvarez 2009). As a result, the panel recognised the need to promote the multifunctionality in olive oil LAFS, especially in areas of traditional olive orchards located on sloping land, through local development strategies that respect the environment and that valorise both the landscape and the cultural heritage. This tallies with the relatively high score attributed by the panel to *scarce efforts to add value to the environmental, social and cultural roles of the olive oil chain* (MULTIF, μ=3.97).

The last problem, although registering a median of 4, was the lowest ranked of the five issues: *lack of innovation in regards to machinery adapted to traditional olive groves on sloping land* (MAQUIN, μ=3.59). The panel made special mention of the fact that it is not currently possible to automate harvesting using tree-trunk shakers for olive groves located on steep slopes, given the structural weakness of these soils and the accessibility problems posed by this type of orchards. Likewise, current machinery is not suitably adapted to vegetal cover control tasks for olive groves in mountainous areas.

3.4 Environmental issues linked to the development of super-intensive olive production

The *widespread verticillium wilt amongst irrigated olive orchards* (VERTICIL; μ=3.94 & Me=4) is a high priority issue that has been particularly affected by the growing intensification of irrigation used in olive farming, particularly in super-intensive olive groves. The panel agreed to cataloguing verticillium as an important issue, but also on the need to apply safeguard measures aimed at repressing, or at least reducing, the spreading of the infectious agent to olive farms.

However, two relatively opposing opinions emerged in regards to how to overcome this disease, as qualitative information showed. One group of experts suggested that this primarily represents a research and investigation issue. They justify the need to prioritise interdisciplinary research programmes that aim to improve specific knowledge concerning mechanisms that spread the disease in each area: polluted water used in irrigation, crops sewn prior to the planting, absence of sanitary certification of the nurseries, etc. On the other hand, the second group stated that the investment in investigating verticillium, which is closely linked to inappropriate farming practices and to intensive production processes, is detrimental to research into other pests and diseases. They believe that research should be directed at biological control, with a view to isolating new natural insecticide molecules, at evaluating potentially infected soil risks or at using resistance rootstocks.

This partial dissent between groups of different environmental stances was also visible in the importance given, only "above average" (Me=3.5), to the *appearance of new environmental problems as a result of super-intensive olive production* (SUPERINT; μ=3.47). The dilemma is based on whether high density groves foster environmental problems, other than the increase of water consumption: very high use of inputs as agrochemicals, drastic reduction in olives varieties or increase of pests and diseases. As little more than ten years have passed since the initial implementation of super-intensive orchards in Spain, environmental effects are still not entirely measurable.

4. Environmental and territorial research and innovation programmes for the olive oil sector

The panel was provided with a list of seven research and innovation programmes to be incorporated, from an environmental, territorial and rural development perspective, into the National R&D&I System.[1] In Table 2-2 and Figure 2-3 the arithmetic means and the medians attributed by the experts for each of the programmes can be seen. The nature of this valuation process was threefold: i) importance of the research and/or innovation programmes; ii) level of certainty in regards to the research results; iii) importance of the result and knowledge transfer actions.

Firstly, many of the experts stated that national research programmes must address urgently cross-interdisciplinary focus, rather than solely individual aspects. They also remarked that these programmes should promote working together various research groups as, to date, cooperation levels among them have been too low, which in turn has resulted in reduced efficiency in overcoming important environmental and territorial issues such as soil erosion, use of irrigation, verticillium management, multifunctionality, etc.

Table 2-2 Environmental and territorial R&D&I programmes for the olive oil sector

Median score of "importance of research and innovation lines": Me I R&I
Median score of "certainty in research results": Me CERT
Median score of "importance of transfer activities": Me I TR

	Me I R&I	**Me CERT**	**Me I TR**	**Abbreviation**
Fighting soil erosion and degradation	**5**	**4**	**5**	**SUELO**
Sustainability of water use and irrigation efficiency	**4**	**4**	**5**	**RIEGO**
Fighting pests and diseases	**4**	**3**	**5**	**ENF-PLAGAS**
Certification systems linked to the territory (designations of origin, organic agriculture, integrated production, etc.): differentiation-related incomes, local governance and rural development	**4**	**3**	**4**	**DO-AECOL**
Multifunctionality and environmental externalities in olive groves and olive millsl	**4**	**3.50**	**5**	**MULTIF**
Value-added strategies for the olive oil natural and cultural heritage. Oleotourism	**4**	**4**	**4**	**PATRIM**
Treatment of biomass and waste water from the olive oil chain	**4**	**4**	**5**	**BIOMASA**

Me I R&I and Me I TR:
1= very little importance; 2 = little importance; 3 = average importance; 4 = high importance; 5= very high importance.

Me CERT:
1= very little certainty; 2 = little certainty; 3 = average certainty; 4 = high certainty; 5= very high certainty

All seven *R&I lines* have been attributed a median of 4, apart from *fighting soil erosion and degradation*, which was scored as a 5. Likewise, the arithmetic means awarded to the research and innovation lines fall within a high, limited range of between 3.91 and 4.43. The methodology sequence used to select and prioritise the corresponding lines and programmes in the two previous phases of the project has meant that all lines and programmes are considered as at least high importance.

The score attributed to *transfer activities*, varying between 3.97 and 4.83, is significantly higher than the scores attributed to R&I lines (figure 2-3) for programmes addressing the *correction of agro-environmental externalities* (SUELO, RIEGO, ENF-PLAGAS and BIOMASA). Furthermore, five of the seven programmes have been scored as of "very high importance" (the four mentioned above and MULTIF). This evaluation can be explained, in part, by the lack of public training and knowledge transfer systems offered to farmers over the last decades.

In regards to the *multifunctionality and olive oil LAFS programmes*, similar scores have been attributed both to transfer activities and to R&I lines (MULTIF, DO-AECOL and PATRIM). Significant work is still required in order to promote interdisciplinary investigation in these areas, due to the fact that this is still a relatively new research area, in addition to the high levels of environmental and territorial variability inherent to the Spanish olive oil LAFS.

The scores awarded to the *level of certainty* by the panel are significantly lower than those attributed to transfer activity and to the importance of R&I lines. Four of the seven programmes were awarded a median of 4 ("high certainty"), whilst three were considered as of "average certainty" (ENF-PLAGAS and DO-AECOL, Me=3) or "above average certainty" (MULTIF, Me=3.5).

Next, the *importance of research and innovation lines* will be analysed and compared to the scores awarded to the *environmental and territorial problems* discussed in Section III.

The first category includes programmes that aim to *correct agro-environmental externalities.* Although the general objectives are based on the reduction of negative externalities, in certain cases, such as olive oil waste, promotion of positive externalities policies can also be important. The average R&I line scores fall between 4 and 4.43, whilst transfer actions scoring even higher, between 4.44 and 4.83:

Fig. 2-3 Environmental and territorial R&D&I programmes for the Spanish olive oil sector

Arithmetic mean score

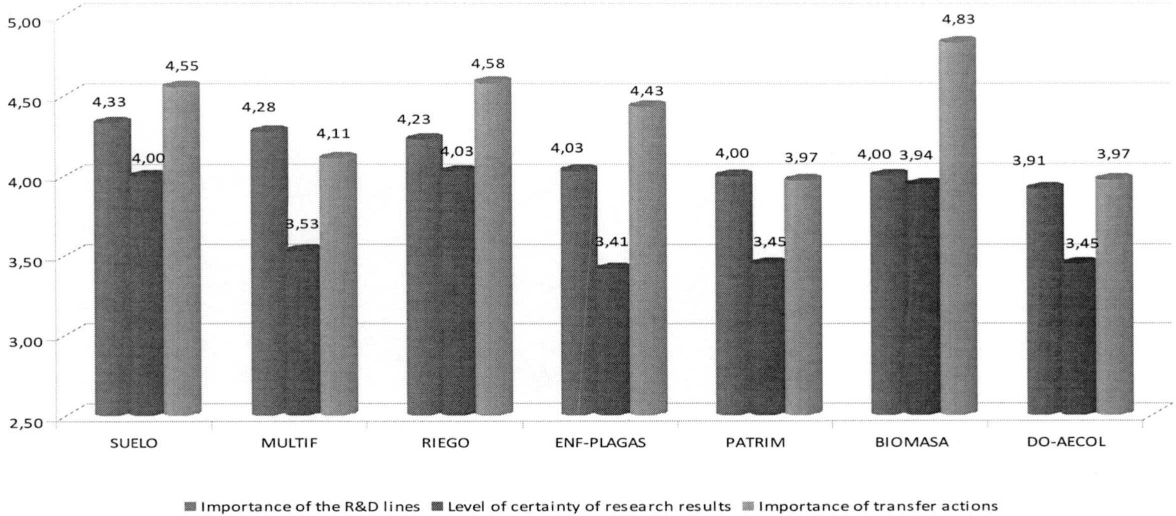

I+D and I TR
1= very little importance; 2 = little importance; 3 = average importance;4 = above average importance; 5= high importance.
L CERT
1= very little certainty; 2 = little certainty; 3 = average certainty;4 = above average certainty; 5= very high certainty

- The highest R&I mark awarded was in the program of *fighting soil erosion and degradation* (SUELO, μ=4.43). Additionally, this is the only programme that was awarded a score of 5 in both R&I and transfer actions. This score is mirrored by the high score awarded to "high level of soil erosion and degradation" problems (EROSION; μ=4.69, Me=5), which is higher than scores of all other thirteen problems.
- The *sustainability of water use and irrigation efficiency* was attributed the second/third highest score of all seven programmes in regards to R&I lines (RIEGO, μ=4.29 & Me=4). Its order of magnitude is similar to that of the following "highly important" problems: "limitations on water resource availability" (HIDR, μ=4.11) and "optimisation of water use in irrigated farms" (RIEGO, μ=3.8).
- The *fighting pests and diseases* programme shares a mean of 4 (ENF-PLAGAS) with problems related to the "excessive use of synthetic phyto-sanitary products" (FITOS) and was scored only slightly higher than "widespread verticillium wilt amongst irrigated olive orchards" (VERTICIL, μ=3.94).
- Finally, one programme aims not just to correct negative externalities, but to also enhance olive oil chain wastes: *treatment of biomass and waste water from the olive oil chain* (BIOMASA, μ=4). Many experts stated that olive biomass should primarily be directed towards restoring soil fertility and overcoming soil degradation rather than the use in energy or bio fuels production.

The second category, *olive oil LAFS and multifunctionality,* encompasses three programmes that all attained relatively high scores: those for R&I lines varied between 3.91 and 4.29, whilst those for transfer actions between 3.97 and 4.21. Together they aim to address problems related to geographical changes and territoriality in olive oil LAFS:

- The score awarded to *multifunctionality and environmental externalities in olive groves and oil mills* (MULTIF, μ=4.29 in R&I), is of particular interest, occupying the second/third place amongst the seven selected programmes. It is generally considered that new methodological frameworks that assess jointly different environmental externalities must be developed urgently in order to design a list of policy priorities. The objective of policies aimed at promoting multifunctionality is to reward economic and social agents in their role of producers of public goods. According to the experts, these policies should consider as a group not only environmental externalities, but socio-economic and cultural ones as well.

- The second and third programmes are linked to environmental and socio-cultural value-added strategies for specific olive oil territorial assets. They integrate private strategies for the commercial goods (oil, cosmetics, touristic goods, etc.) with public policies. *Adding-value strategies for the olive oil natural and cultural heritage; oleo-tourism* (PATRIM, μ=4), address the enhancement of the local olive landscapes and olive oil cultural assets, whether through oleo-tourism strategies, or creating marketing synergies together with the local oil firms and institutions. The *certification systems linked to the territory* (DO-AECOL; μ=3.91) programme looks to foster inter-professional quality organisation strategies for LAFS, aiming to obtain differentiation-related incomes associated with the collective labels and to contribute to the territorial governance processes; this is particularly the case for designations of origin boards and organic agriculture associations,

The assessment of olive oil LAFS and multifunctionality programmes is consistent with the evaluation of main environmental and territorial problems linked to them. These problems registered scores between 3.89 and 4.06: "need to restructure medium and low-yield olive groves" (REESTR), "cost reduction requirements for extensive olive production" (COSTES) and "geographical olive grove changes as a result of increased intensive and super-intensive cultivation" (GEOGR).

5. Conclusions

The research project, aiming to establish research and innovation priorities in the Spanish olive oil sector, verified the adequate application of a combination of social research techniques to expert panels, in order to resolve complex problems in the area of environment, territory and local olive oil development. Delphi methodologies facilitated, on one hand, the analysis and valuation of R&D&I programs and lines and, on the other hand, olive oil LAFS environmental and territorial problems. In this study, the panel used a cross-interdisciplinary perspective, integrating environmental, agricultural and socio-economic approaches. Additionally, it is also necessary to develop programmes on a national scale that overcome excessive research group fragmentation, as is common place in the case of multifunctionality and verticillium, among other topics.

Another conclusion of the study is the urgency with which the Spanish R&D&I system must foster transfer programmes. In many of these knowledge areas, transfer activities require more immediate attention than the promotion of strictly research lines; this is particularly apparent in regards to soil degradation and sustainable use of water. The adoption of applied research and transfer programmes must consider agro-environmental heterogeneity, as well as the diverse farming, agro-industrial and management practices that are present in the Spanish olive oil LAFS. A truly territorialised focus must be adopted.

Additionally, the study has facilitated the establishment of seven R&D&I programmes, grouped under two different categories: firstly, agro-environmental externalities corrective actions and, secondly, actions that address olive oil LAFS and multifunctionality. These programmes are relatively harmonised, in terms of their relative importance, with the main environmental and territorial problem categories that Spanish olive oil LAFS must overcome.

With regards to the first category, the high level of agro-environmental externalities produced by the Spanish olive oil LAFS is principally visible in high soil erosion and degradation. The panel identified this as the most important territorial problem facing the Spanish olive oil sector; the respective research and transfer actions have also been clearly attributed the highest priority. Correcting soil degradation forms the basis of the recommendations that interacts with other agro-environmental externalities that have a significant impact on the Spanish olive oil sector, such as diffuse pollution, loss of biodiversity, lack of organic fertilisation, water shortage and the need to treat olive oil waste.

Secondly, olive oil LAFS and multifunctionality R&DI programmes and lines aim to respond to the enhancement of public and commercial goods. The key issue is to reduce/promote the negative/positive territorial externalities linked to olive oil local development strategies. These proposals address olive oil LAFS problems relating to olive grove geography, such as the survival of marginal olive groves, restructuring of low and medium-yield olive groves and the environmental impacts of olive farming intensification. In addition, it is necessary to solve the problem of lack of methodologies aimed at assessing how to distribute the reward of public goods to economic and social agents. Multifunctionality policies must adopt a clearly territorialised focus.

In this sense, the panel agreed that urgent actions are required in order to promote multifunctional public policies directed at avoiding the abandonment of mountain olive groves. It must be noted that, to date, financial support provided to traditional olive grove systems likely to generate positive environmental externalities and local olive sector development has been lower than the support received by the Spanish olive sector from direct payment policies and market support. It is hoped that, in light of the 2013 reforms, significant changes will be made that favour the rewarding of environmental, socio-economic and cultural goods. Organic olive production, designations of origin, locally-driven joint marketing, natural and cultural heritage enhancement strategies, in addition to a series of networks and institutions resulting from collective action, all have the potential to play a lead role in olive oil territorial development in the near future.

Notes

This study has been financed by 2008-2011 National Plan R&D&I research projects. Spanish Ministry of Science and Innovation: i) EXTERSIAL: Ref. CSO2009-08154 (2010/12). Fundamental Research Program. Principal Investigator: Dr. Javier Sanz Cañada. ii) Project RET-060000-2010-4 (2010/11), co-financed by the ERDF: ALENTA, Olive Oil Technological Platform. National Network Program. Coordination: Citoliva Foundation.

1. Each of these seven programmes groups different research and innovation lines, 37 in total, which were both evaluated and prioritised by the panel. A detailed analysis will be the subject of a future study.

References

AEMO (2010). *Aproximación a los costes del cultivo del olivo*. Córdoba: Asociación Española de Municipios del Olivo.

Arriaza, M. and Nekhay, O. (2010). Evaluación social multicriterio del territorio agrícola: el caso del olivar de baja producción. *Revista de Estudios Agrosociales y Pesqueros* 226: 39-69.

Cairol, D. et al. (2009). Multifunctionality of agriculture and rural areas as reflected in policies: the importance and relevance of the territorial view. *Journal of Environmental Policy & Planning* 11 (4): 269-368.

Calatrava, J., Franco, J.A. and González Roa, M.C. (2007). Analysis of the adoption of soil conservation practices in olive groves: the case of mountainous areas in southern Spain. *Spanish Journal of Agricultural Research* 5 (3): 249-258.

Economies et Sociétés (Comp.) (2007). Dossier Systèmes Agroalimentaires Localisés. Économies et Sociétés. *Systèmes Agroalimentaires* 9: 1463-1565.

Fleskens, L. and Graaff, J. de (eds.) (2008). A sustainable future for olive production on sloping land. *Journal of Environmental Management* 89 (2), special issue: 73-142.

Francia, J.R., Durán, V.H. and Martínez-Raya, A. (2006). Environmental impact from mountainous olive orchards under different soil management systems. *Science of the Total Environment* 358: 46-60.

García-Azcárate, T., Martin, H. and Orlandi, B. (2010). *Life among the olives: good practice in improving environmental performance in the olive oil sector*. Brussels: European Commission Life Focus.

Gómez-Calero, J. (ed.) (2009). *Sostenibilidad de la producción de olivar en Andalucía*. Sevilla: Junta de Andalucía.

Graaff, J. de et al. (2010). The future of olive groves on sloping land and ex-ante assessment of cross compliance for erosion control. *Land Use Policy* 27: 33-41.

Guzmán-Alvarez, J.R. (2004). *Geografía de los paisajes del olivar andaluz*. Sevilla: Junta de Andalucía.

Guzmán-Álvarez, J.R. and Navarro, R.M. (2008). Modelling potential abandonment and natural restoration of marginal olive groves in Andalusia (south of Spain). *Journal of Land Use Science* 3 (2): 113-129.

Mann, S. and Wüstemann, H. (2008). Multifunctionality and a new focus on externalities. *The Journal of Socio-Economics* 37: 293-307.

MARM (2009). *Encuesta sobre superficies y rendimientos de cultivos*. Madrid: Ministerio de Medio Ambiente, Medio Rural y Medio Marino.

Muchnik, J. and Sainte-Marie, C. de (eds.) (2010). *Le temps des SYAL*. Paris: Editions Quae.

Muchnik, J., Sanz-Cañada, J. and Torres-Salcido, G. (2008). Systèmes Agroalimentaires Localisés: état de recherches et perspectives. *Cahiers d'Études et des Recherches Francophones/Agricultures*, 17 (6): 513-519.

Nekhay, O., Arriaza, M. and Guzmán-Álvarez, J.R. (2009). Spatial analysis of the suitability of olive plantations for wildlife habitat restoration. *Computers and Electronics in Agriculture* 65: 49-64.

Oakley, P. (1991). *Projects with people: the practice of participation in rural development*. Geneva: International Labour Organization.

Renting, H. et al. (2009). Exploring multifunctional agriculture: a review of conceptual approaches and prospects for an integrative transitional framework. *Journal of Environmental Management* 90: 112-123.

Sanz-Cañada, J. (2009). Les appellations d'origine protégée d'huile d'olive en Espagne: systèmes agroalimentaires localisés, gouvernance et externalités territoriales. In *Les produits de terroir, les indications géographiques et le développement local durable des pays méditerranéens*, edited by Tekelioglu, Y., Ilbert, H., Tozanli, S., 237-254. Montpellier: CIHEAM.

Sanz-Cañada, J. and Macías, A. (2005). Quality certification, institutions and innovation in local agro-food systems: protected designations of origin of olive oil in Spain. *Journal of Rural Studies* 21: 475-486.

Sanz-Cañada, J. and Muchnik, J. (eds.) (2011). Ancrage et identité territoriale des Systèmes Agroalimentaires Localisés. *Économie Rurale*, 322 (dossier): 4-83.

Sanz-Cañada, J. et al. (2008). *Prioridades de investigación y innovación en el sector del aceite de oliva en España*: *Informe de la Red de Cooperación Ciencia y Empresa del sector oleícola. OLIRED*. Madrid. http://hdl.handle.net/10261/11141.

Stobbelaar, D.J. et al. (2009). Multifunctional agriculture: from farm diagnosis to farm design and institutional innovation. *Journal of Environmental Management* 90: 109-111.

Zander, P. et al. (2007). Multifunctionality of agriculture: tools and methods for impact assessment and valuation. *Agriculture, Ecosystems and Environment* 120: 1-4.

CHAPTER THREE

VALUE CHAINS AND CHAINS OF VALUES: TRACING TANZANIAN TEA

ALLISON LOCONTO

Abstract

The objective of this chapter is to examine value chain governance through case studies of four different certified value chains for Tanzanian tea. This chapter takes a look at a traditional export commodity, tea, and discusses the implications of involvement in value-based certification schemes (Ethical Trading Initiative, Fairtrade, Organic and Rainforest Alliance) on certified producers in Tanzania. This chapter specifically analyses the network construction of each certified value chain and answers the questions: 1) which actors are involved in each value chain, 2) which values are claimed as organizing principles of these value chains, and 3) what does this mix of actors and values contribute to our understanding of value chain governance. Between 2008 and 2010, seventy-eight in-depth interviews were conducted with stakeholders who are certified against sustainability standards. Twenty focus groups were formed comprising certified smallholders and hired labourers. The conclusions suggest that despite claims about the ability to change trading relationships by creating certified value chains, most of the old networks are still in place. The certification systems only add additional buyers to global value chains that were already governed by highly relational and hierarchical mechanisms. These conclusions thus call into question some of the claims made by certification schemes as to their abilities to change practices.

1. Introduction

Changes in the global food regime have brought increasing pressure on African producers as they are being included in global agri-food value chains in both old and innovative ways. This focus on a global value chain has increased support for export driven agriculture, while diversifying the nature of crops cultivated for export. The traditional commodities of coffee, cotton, cocoa, and tea are still being promoted through these systems, while new crops such as horticulture and cut flowers are becoming more popular. With these new crops and the use of a supply chain management (SCM) approach to global sourcing, some of the value-added practices such as processing and packaging are also moving into the African agricultural landscape. Moreover, the shift towards retailer-driven global supply chains for agri-food products and the provisions made to protect against Technical Barriers to Trade and Food Safety in the World Trade Organization agreements have increased attention to the use of standards to regulate quality and production around the world (Loconto and Busch 2010). As a whole, these practices are changing the shape of African agriculture for those who can afford to be included in the process. Who then is included in this process and who is excluded?

Within these broader trends in the global agri-food system, this chapter examines the traditional export commodity of tea. Tea has been grown commercially in Tanzania since the 1930s, first on colonial plantations and second by smallholders beginning in the 1960s. In the case of tea, the first stage of processing green leaf tea into dried made tea (MT) has traditionally been located close to the fields in Africa. Thus the recent shifts in processing locations are not new to the tea sector. However, the use of SCM techniques by large buyers and supermarkets are influencing the tea trade. Specifically, certification for sustainability standards in tea production has recently received substantial attention in the industry as the largest international tea buyers (e.g., Lipton and Tetley) are making certified tea a mandatory requirement for inclusion in their value chain.

This chapter discusses the governance trends in value-based certification schemes, namely the Ethical Trading Initiative (ETI), Fairtrade, Organic and Rainforest Alliance (RA). Each of these certification schemes makes claims on specific values that are embodied by a particular "value chain". This chapter specifically analyses the network construction of each certified value chain and answers the questions: 1) which actors are involved in each value chain, 2) which values are claimed as organizing principles of these value chains, and 3) what does this mix of actors and values contribute to our understanding of value chain governance? The remainder of the chapter follows in four sections. The conceptual framework of Global Value Chains (GVC) analysis that is used in this chapter is introduced and the traditional value chain for Tanzanian tea is explained. The following section traces the four value chains that are emerging within the Tanzanian tea sector based on the qualities of certified production. The section analyses, from a global value chains perspective, which actors are included in these values-based value chains, the values they claim to promote, and the governance implications of this involvement. Broader implications conclude.

2. Conceptual framework: value chains

Global Value Chains (GVC) analysis proposes that research on value chains follow a product from field to cup (Gereffi et al. 2005). GVC analysis has emerged from research in the 1980s on commodity systems (Bair 2009; Friedland 1984), which consist of the value chain that converts a natural resource into a consumer item and the supplementary activities that support that value chain. A value chain typically has about five stages: production, processing, distribution, retail, and consumption. Recent research has shown that rather than static vertical chains, value chains are webs of interaction, where negotiations take place between actors at each node (Gibbon and Ponte 2005; Tallontire 2007).

Gereffi et al. (2005) identify five types of GVC governance–hierarchy, captive, relational, modular, and market–ranging from high to low levels of power asymmetry. These types of governance are differentiated according to: (1) complexity of inter-firm transactions; (2) the degree to which this complexity can be mitigated through codification; and (3) the extent to which suppliers have the necessary capabilities to meet the buyers' requirements (Gereffi et al. 2005). There are three main trends in GVC governance: 1) *Producer-driven*, where MNCs or large integrated industrial enterprises play the central role in controlling the production system. Downstream traders and retailers are coordinated by a large producer (e.g., car dealerships), while low-profit activities are outsourced upstream to networks of suppliers. Contracts bind suppliers to strict product and process standards (Gereffi and Korzeniewicz 1994); 2) *Buyer-driven*, with high barriers to entry and high profits located at the retail end with the protection and promotion of brands (Conroy 2007; Gereffi et al. 2005; Gereffi and Korzeniewicz 1994); or where large retail establishments use their market power to effectively govern the production of commodities with the desired attributes (Fold and Larsen 2008); and 3) *Twin-driven*, where lead firms govern the supply network, while environmental groups/movements and third-party certifiers/standards developers govern the regulatory aspects of the network (Islam 2008).

While these typologies focus on economic power within the organisation of value chains, cultural and institutional power are also vital to their governance (Dixon 1999; Freidberg 2004; Neilson and Pritchard 2009), particularly with respect to notions about quality (Gibbon and Ponte 2005; Ponte and Gibbon 2005) and stakeholder preference (cf. Ochieng 2008; Pirsch et al. 2007). I utilize these notions to show how standards and certification are used to facilitate transactions and to discipline the people, processes and products in the value chain (cf. Loconto 2010).

Data was collected for this study in Germany, Kenya, Tanzania and the United Kingdom between 2008 and 2010 and consisted of document analysis, in-depth interviews (78), focus groups (20) and observations. Qualitative analysis was conducted using NVivo8 software. The presentation of the case studies identifies the actors involved in each value chain between production and packing. Governance relations are identified and power relations in terms of "drivenness" are explored. First I present the traditional value chain for tea as a baseline for analysis.

Figure 3-1 Generic Tea Value Chain

3. Description of the traditional tea value chain

Tea (black, green and white) is made from the leaves of an evergreen bush (*Camellia sinensis*) which thrives at high altitudes (1400-2500 m) in the humid tropics and sub tropics. All factories in Tanzania use the cut-tear-curl (CTC) method of tea processing for the tea sachet market. Orthodox tea is a higher value market and consists of gently rolling rather than cutting and tearing the tea prior to the fermentation process. This process is also used in two of the Fairtrade and Organic certified factories.

The tea value chain has traditionally been a vertically integrated, hierarchically governed chain (Figure 3-1). Tea has a year-round harvest where each field is plucked every 7-20 days depending on the season. The green leaf collected from the fields must reach the processing factory in less than 12 hours, which means there are very small catchment areas around factories where both estates and outgrowers can grow tea (Simbua 2006). This creates a situation of close relationships between factories and farmers and limits the emergence of competitive markets for green leaf. These close relationships have long been formalised into contract farming schemes, which provide inputs on credit to outgrowers (FAO 2009).

Outgrowers are paid on a monthly basis for the delivery of green leaf at the average rate of US$ 0.10/kg in 2009.

A tea processing factory is considered to be the anchor of the tea value chain in Tanzania, where the most local decision-making power lies (Simbua 2006). Most investors have purchased processing factories together with the purchase of large-scale estates. In Tanzania, the tea packing companies have also invested upstream through joint investments in made tea (MT) processing factories and/or estates that provide the majority of the MT that is packed domestically (e.g., Afritea & Coffee Blenders have a joint interest in the Lushoto Tea Company which owns the New Mponde Factory in the Usambaras). In 2010, there were 19 MT processing factories in Tanzania,[1] 18 of which had estates and one that was sourced solely by smallholder farmers.

Market prices are determined every Monday and Tuesday at the Mombasa Auction in Kenya, although a majority of Tanzanian tea does not pass through the auction and is instead sold through private contracts (Baffes 2003; TBT 2009). Tea quality plays a vital role in determining the final value at auction. The assessment of quality (through tasting by brokers and buyers) distinguishes the value of tea across different factories and this allocation of quality has legacy effects for the reputation of the tea. At the Auction, brokers sell tea on behalf of the producers. This tea is bought by the first set of buyers (blenders or international traders) who purchase lots of 40-60 packages of tea that have an average weight of 60 kg. On the auction floor, these lots can be split during the bidding process so to stop endless bidding (and thus price increases) on the same lots.

These buyers blend the single estate teas into what they call a "standard" and ship this to their international buyer, who put in their orders for tea from specific estates and/or exhibiting specific quality attributes prior to the weekly auction. This international buyer is known as a tea blender or tea packer. The main destinations for tea from Tanzania are Egypt, UK, Pakistan, and Sudan (TBT 2009). At this stage the "standard" blend bought in Mombasa is blended again with tea bought from auctions around the world and is packed for consumption. This packed tea is then sold to a variety of retailers who sell to consumers.

4. Chains of value

In this section, the four value chains are traced introducing the actors involved at the different points of value addition outlined above, as well as those intervening in the broader institutional context of each value chain.

4.1 The Ethical Trading Initiative (ETI)

The role of the ETI standard in Tanzanian tea value chains takes two main forms, first in the form of Typhoo Tea and second through the Ethical Tea Partnership (ETP) which is that most discussed in the industry. Both ETP and Typhoo are members of the ETI and run a "monitoring" programme based on the ETI base standard (Blowfield 2004).

Typhoo Tea Limited is a UK-based blender that has blending and packing factories near Liverpool in the UK. It was established in 1903 and since then has grown to sell their products in over 40 countries including USA, Canada, Australia, and South Africa. Typhoo has a number of diverse and successful brands such as Typhoo speciality tea and fruit infusions, London Fruit & Herb, Ridgways Tea, Heath & Heather, Melrose's Tea, QT Instant Tea, Lift Instant Tea, Brook Bond D Tea, and Fresh Brew Tea. The Typhoo factory is also linked with Global Tea and Commodities Limited (GTC), which is another UK-owned agribusiness that is vertically integrated by owning estates, MT processing factories and blending and packing factories in East Africa under the name of Gold Crown Foods (Epz) Limited. The Typhoo operations in Liverpool blend and pack tea for GTC contracts amongst others including supermarket brands in the UK. At the auction GTC acts as the first buyer for Typhoo. For Tanzanian tea, Typhoo buys directly from the producer factory and utilizes the traditional value chain. Typhoo claims to "have a long tradition of working closely with [their] suppliers," which enabled them to launch the Quality Assurance Programme (QAP) in 1992 (Typhoo 2010, 1). The QAP self-assessment questionnaire also asks if the producer is certified against the HACCP, Fairtrade, RA and UTZ Certified standards. The quality assurance manager at Typhoo will typically also make a field visit; however, multiple certifications make a field monitoring visit redundant.

In 1997 "a number of large tea companies made the ground-breaking decision to work together to promote sustainability in the sector, rather than competing on ethics" (ETP 2010, 1). This was the beginning of the ETP and membership includes 20 international tea packers (e.g., Tetley, Twinings, The Republic of Tea, TAZO, Taylors of Harrogate and Sara Lee) with markets in Europe, North America, Australia and New Zealand. Founding members Unilever and James Finlay have left the partnership since 2007 and have opted for RA and Fairtrade standards.

The ETP tries to improve five areas of tea production labour rights, education, maternity, health and safety, housing and basic rights, and the environment (ETP 2009). The mission of ETP is "to make this picture transparent—to monitor living and working conditions on tea estates, with the aim of making sure that the tea you buy from the members of our Partnership has been produced in a socially responsible way" (ETP 2009). The members of the ETP are the UK based tea blenders/packers and monitoring (or verification audits) of the producing estates and factories, paid by the members, serve to ensure that the mission of the ETP is being achieved. Within the value chain, the UK blenders (buyers) request that producers complete a self-assessment questionnaire provided by the ETP. Non-conformances are identified through ETP's review of the self-assessments and a field monitoring visit is supposed to occur every three years. In 2009 ETP switched its auditor for the scheme to AfriCert Ltd., a Nairobi based agricultural certification agency that is also accredited for GLOBALGAP, BRC and is contracted for the RA audits.

In 2010, 74 percent of the MT processing factories in Tanzania were involved in GVCs and all 100 percent of these have been monitored by ETP during the past ten years. According to the standard, factories should lose certification if they exceed the maximum level of non-conformances, but in practice this is not the case (Interview July, 2008). This renders the value of the ETP monitoring visit to little more than an informational visit as ETP's members continue to rely on the traditional value chain to purchase tea from their preferred producers.

The ETP appears to be moving into a "capacity building" role within the tea sector (ETP 2009). The ETP plans to assist producers in becoming certified against RA, Fairtrade and UTZ certifications. The focus is thus shifting from monitoring non-compliances to identifying core issues in the industry and working with producers to eliminate them. ETP is beginning to collaborate with the ISEAL (International Social and Environmental Accreditation and Labelling) Alliance, which is an apex organization.

4.2 Fairtrade

In the context of Tanzanian tea, Fairtrade[2] refers to the Fairtrade Labelling Organizations International (FLO) standards and certification system which was officially established in 1997 to create cohesion amongst existing fair trade marks in Europe and North America. FLO standards define the rules for those who are governed within their value chains and are based on the ideas of "transparency, partnership and participation, representative democracy, and equal exchange" (FLO 2010). Within the Fairtrade system certification is needed for producers (farms and factories), buyers (international traders) and blenders/packers. Producers are evaluated according specific "social, economic and environmental development indicators." Both estates and smallholders can become certified for tea.[3] Fairtrade provides an additional monetary premium to producers and varies according to the product (CTC or Orthodox) and the type of producer, Hired Labour company (HL) or Small Producer Organisation (SPO), but is about 0.50 $/Kg of MT. Estate companies are allowed to keep 0.10 $/Kg out of this premium as a "sustainability margin" so to make workplace improvements to the factory and estate (cf. Loconto and Simbua 2012).

The value chains for fair trade are numerous and there are a large number of different actors involved at each level. Fair trade certified tea in Tanzania began in 1994 with Mufindi Tea Company's (MTC) factories: Luponde and Itona. This was characterized by research participants as an 'ethics' based decision (Interview October, 2009). There are currently five factories certified against the Fairtrade standards – four ow which are certified via the HL mechanism, while the fifth sources from a certified SPO.

In 2009, the tea market report showed that Fairtrade certified tea accounted for 7 percent of the UK's tea sales in 2009 with the top three retailers being Tesco (market share 31%), Asda (17%) and Sainsbury (16%) (van Reenen, Panhuysen, and Weiligmann 2010). Supermarket own brand teas have been a large part of the market increase in Fairtrade certitied tea with Sainsbury, Co-op and Marks & Spencer launching public commitments to Fairtrade (van Reenen, Panhuysen, and Weiligmann 2010). However, this large increase in market demand has not translated into significant increases for Tanzanian tea, in 2009, the percentage of total sales for the Tanzanian Fairtrade sector was an average of eight percent.

Fairtrade encourages the use of the traditional value chain relations via their provision for retroactive certification of bought tea.[4] "In 2008, systematic retro-certification of Fairtrade teas was phased out to overcome a lack of transparency and "unfairness" associated with the practice and perceived inconsistency with Fairtrade trade rules" (FLO 2009, 12). However, the 2010 revision of the Tea standard reintroduced retro-certification across the board (FLO 2010). This provision has enabled many of the large multi-nationals and UK supermarkets to continue their practice of purchasing tea through the auction system (cf. Dolan 2008).

Cafèdirect has been purchasing tea directly from two FLO certified factories in Tanzania (Kibena and Katumba) since 1998 and 2003 respectively. These two factories are also shareholders in Cafèdirect and are involved in a number of additional "development" projects in collaboration with Cafèdirect. As one of my research participants explained: "our business model definitely goes much beyond the FT minimum!! [...] What I would like to stress however in line with above that this initiative (as a public-private-partnership with GTZ/ Germany)[5] is precisely NOT Fairtrade Value Chain but a specific Cafèdirect value chain and understanding of what a responsible & ethical business should do nowadays" (email from research participant, participant's own emphasis).

The case of the German market differs from the UK market both in terms of scale and in terms of the relationships with Fairtrade producers. In 2009, Fairtrade tea sales grew by 13 percent from 2008 to 229 tonnes of tea (which represents about one percent of the tea market), 77 percent of which were also certified Organic (TransFair 2010). Critical to this step forward was the increase of Fairtrade certified tea packers: Cha Dô, GEPA, Teekanne, Tee Geschwendner and Rossmann. The specific use of Tanzanian tea is to add this tea to its blends and is based purely on the purchase of tea from the two organic certified factories: Herkulu and Luponde (an MTC factory). As such, only Cha Dô and Kirchner, Fischer & Co. GmbH (KFCo) are involved in the Tanzanian Fairtrade value chain (further explanation follows in the next section). Cha Dô is highly engaged in the FLO system and sits on Product Advisory Council.

There is also a value chain that is part of a fair trade value chain but is not part of the certified Fairtrade value chain. Specifically, the Fairtrade certified trader, El Puente GmbH, is considered to be an Alternative Trade Organisation (ATO) and buys non-certified packed tea from a Tanzanian tea blender and packer, Afri-Tea and Coffee Blenders (1963) Ltd. (Chai TTB), which sources its tea directly from its own MT processing factories that are supplied directly from smallholder farmers. These two factories (Mponde and Lupembe) are not Fairtrade certified, but the smallholder associations own 50 percent of the factory. In addition to paying premium prices for the packed tea, El Puente also contributed funds to projects meant to help rehabilitate the Mponde factory and improve the tea production of the smallholder farmers.

We also see a large number of NGOs and governmental organisations involved in providing policy and capacity building support for Fairtrade. Many of these initiatives take place at the growing level of the value chain and consist of commercial extension services by the Tea Research Institute of Tanzania, awareness-raising by Technoserve, capacity building by Imani Development and Technoserve and "liaison" services by FLO. At the policy level FLO receives support from GTZ and the UK Department for International Development (DFID) in terms of public-private partnership programmes at the producer level. For example, GTZ helped Cafèdirect fund the three-year long AdapCC pilot projects that brought together Cafèdirect's coffee and tea producer members from around the world to conduct risk assessments for adaptation to Climate Change. FLO originally worked in partnership with the World Fair Trade Organizations (WFTO), but this relationship has been described recently as a contested as they are now seen by WFTO members as a competing fair trade systems in the struggle to define what "fair trade" means in the market (Interview May, 2010). FLO is a full member of the ISEAL Alliance.

4.4 Organic

The International Forum for Organic Agriculture Movments (IFOAM) principles claim that the basis for organic agriculture is an ethic of "health, ecology, fairness and care" (IFOAM 2010). This holistic system based on agro-ecology has been incorporated into law in 74 countries around the world. While there is an East African Organic Standard, those that are currently applied in the Tanzanian tea sector are the EU regulation (EEC 2092/91) and the US National Organic Program (NOP). The organic certification is also the only certification that covers all stages of the value chain (through distribution).

In Tanzania the two certified organic estates are also certified Fairtrade. The move to organic was made before the Fairtrade certification as both the organic certified companies claimed that they first joined the system because they had acquired overgrown fields and the costs for conventional rehabilitation were much higher compared to the organic certification (Interviews November, 2009 and January, 2010). There is debate as to which of the two estates was the first to be organic certified. Herkulu Estate, owned by Bombay Burmah Trading Corporation Ltd., claims to have been organically cultivated since 1989 and earned its first certificate in 1994 from the Institute of Marketecology, Switzerland (IMO) (Bombay Burmah 2010). Luponde, owned by MTC, claims to have received its first organic certification in 1988 from the Soil Association (Interview November, 2009) and is currently certified organic through Ecocert, Madagascar. The GVC is used in organic to communicate the preference of a specific certification body (and certification mark) by the international buyers (blenders) to the factory management, who then must hire the certifier. Organic tea is currently only bought through direct contracts and accounted for an average of 25 percent of total sales in 2009. Organic tea receives a price premium. The amount of which is difficult to determine because the majority of exported certified organic tea is orthodox processed, which carries an additional price premium.

"Germany is the largest organic food market in Europe and Organic tea accounts for about 4 percent of the German tea market" (van Reenen,

Panhuysen, and Weiligmann 2010, 1), and as mentioned above, 77 percent of the Fairtrade certified teas are cross-certified with Organic. German blenders are also very active in promoting the Tanzanian organic tea industry. Cha Dô is the largest organic and Fairtrade specialty blender in Germany and supplies blended teas to the UK and North American marketsCha Dô buys exclusively from Herkulu in Tanzania to mix with organic tea from other producing regions and is a sales agent of the company that markets Herkulu's tea. In addition to purchasing under both the organic and Fairtrade labels, Cha Dô provides technical assistance and consulting on how to improve the quality of the tea and the organic practices (Interview March, 2010). KFCo owns the brand Mt. Everest Tea Company, one of the oldest specialty tea companies in Germany and exporter to Europe and North America. Specializing in orthodox blends, KFCo only began purchasing Luponde teas in 2010. They were first introduced to it through a taste-test organized by the UK based wholesaler Thompson, Lloyd & Ewert, who has been a long term broker of MTC tea (Interview May, 2010). Therefore, while Cha Dô has vested interests in improving the quality of Herkulu tea, KFCo was first attacted to Luponde tea based on its already existing organoleptic qualities.

Organic orthodox teas from Luponde are also the source of a direct GVC to the UK, where they are sold in an exclusive boutique of the same name in the Burlington Arcade in London. The retail company and the estates are separate companies with a common investor, which has facilitated the development of this GVC based on intimate knowledge of the estates. Currently only the packing and retail of the tea occurs in the UK, but there are plans to pack the tea in Tanzania in the future.

Organic tea is the only certified tea that also appears in a domestic value chain. Chai TTB and Chai Bora, the two leading tea packers in the Tanzanian market, both offer organic brands. The selection includes black tea, green tea, mint, camomile, lemon verbena and hibiscus infusions sourced mostly from Luponde. Organic tea can be found in the large supermarkets in Dar-es-Salaam and Arusha, in the airport shops and in tourist hotels. In these circumstances the estates are mobilizing the investment made for the international market, particularly the efforts made to separate organic from conventional tea, and have exploited them for the local elite.

There are also a large number of institutional actors involved in organic tea, which is illustrative of a highly developed institutional context that supports organic agriculture in Tanzania. Between 1997 and 2008, Sida (Swedish International Development Co-operation Agency) financed a programme to promote exports of organic products from Africa.

Through this programme a number of Tanzanian organisations were created (e.g., TOAM and TANCERT) and a national organic policy was written. These organizations are linked to IFOAM, which was a full member of the ISEAL Alliance until 2010.

4.5 The Rainforest Alliance

The second-largest company in Tanzania, Unilever (PG Tips and Lipton brands), is a key player in the global consumer market with a market share of 12 percent of the global black tea supply in 2009. It is perceived to dominate the trade (Bedford et al. 2002). In 2005, Unilever announced a landscape change in its tea business, promising to make its tea supply chain "sustainable" by 2015 (Unilever 2007). This sustainability is based on the use of the RA standard to verify the sustainability of tea production. In 2010, Tetley, Twinings, and Taylors of Harrogate made similar announcements (RA 2010).

The RA Certified seal confers compliance with the environmental, social and economic standards of the Sustainable Agriculture Network (SAN)–a coalition of conservation organisations originating in Latin America in the early 1990s. While the SAN standard focus on ecosystem conservation, worker rights and safety, wildlife protection, water and soil conservation, agrochemical reduction and education for farm children, the RA network fosters the values of "economy, ecology and ethics" in coordinating its value chain (RA 2010). As part of the process that created tea specific criteria for the SAN, the original 2008 audit of tea plantations in Tanzania was carried out by SAN certifiers. Subsequent audits will be conducted by Africert, who is also responsible for the ETP audits and is an affiliate of the ISEAL Alliance. In Tanzania, to date, the RA standard has only been adopted by Unilever, but interest has been expressed by other growers to try to join the system.

Globally, Unilever operates a vertically integrated value chain where Unilever owns green leaf production, MT processing factories, buying (through direct sales), blending and packing. In Tanzania, this is also the case. Ninety-five percent of the green leaf that the Unilever factories process comes from their own plantations, the other five percent comes from medium-scale growers.While most RA certified tea is currently being bought through direct contracts, there is no preference against the use of the auction system for the RA. In fact, Unilever plans to to continue to rely on the auction system for tea trading, with the hope to make trade in certified tea a normal part of the tea business (Interview December, 2009). To encourage this shift towards sustainability, Lipton (Unilever's buyer) offered a \$0.10/kg premium that was phased out in 2010. This created more interest in the system and other large Tea blenders (notably Tetley and Twinings) have since committed to joining. This proposition points to the need of an increased supply of certified tea, which was not the case in 2010. Indeed, research participants concurred that much investment must be made in order to meet demand for certified tea.

This "demand-driven" approach to certification means that there are interested producers who have been excluded from the system for two main reasons: 1) the lack of RA presence in the region, and 2) RA's policy of attending first to those producers who have an existing demand for their tea (i.e., Lipton, Tetley and Twinings' preferred producers), and then to those who are trying to enter the market. This later group remains the most difficult; as there is a significant need for training and technical assistance in order to comply with the SAN standard. Funding for these programs are external to the cerfiticaiton system and thus the RA has been seeking donors and partners in order to expand the applicability of their standard to small and "non-preferred farmers" (Interview June, 2010).

Alongside the value chain for large blenders, Taylors of Harrogate have also begun to source RA certified tea. Their commitment is linked to their company reputation and their relationship with their tea producers. Taylors of Harrogate is a family-owned company that maintains "trust-based" relationships with their preferred producers. Therefore, the move toward RA certified tea was a product decision based on the decisions made by their producers. In other words, in coffee, Taylors of Harrogate has opted for Fairtrade certification because the majority of their producers were going for that certification, whereas in tea they found that their producers were leaning towards the RA certification. Therefore, in order to maintain their long-term relationships with their producers, Taylors decided to use the RA system (Interview 100603). Taylors continues to purchase their tea directly from their habitual producers.

There are relatively few institutional actors as the RA network is new to the industry. However, RA has signed a memorandum of understanding (MOU) with the ETP, and RA is a full member of the ISEAL Alliance.

5. Discussion

For Gibbon and Ponte (2005), African commodities have been subjected to buyer-driven chains guided by international lead firms. These firms assert monopoly power because of their size and ability to define quality standards. Quality standards pose major barriers to market entry for many African producers who are unable to maintain predictable supplies of the product at the specified standard. Thus, African producers face a buyers' market that demands exacting quality standards, with which they have little direct contact. While African-based producers have persisted in some chains, much of their success is due to private coordination, usually associated with powerful non-African elites. Tea is illustrative of this description of value chains. For example, the Tropical Commodity Coalition claims that "the buying and retailing end of the [tea] market is dominated by a handful of powerful multinational companies that are in a strong negotiation position and are able to skim off most of the value in the chain" (van Reenen, Panhuysen, and Weiligmann 2010).

However, governance also includes those social actors that influence decisions about sustainable production at the various stages of the value chain. These include the many NGOs and other institutional actors that attempt to influence both public and private policies and behaviour concerning sustainable production (Guthman 2009). Some of these NGOs (e.g., Fairtrade) both set standards for sustainable agriculture, and assess and certify compliance with those standards. This shift represents that which is occurring more generally in the global agri-food system: a decrease in regulation and an increase in voluntary forms of governance. A such, these cases are examples of what Islam (2008) calls twin-driven commodity chains or what Tallontire et al. (2011) claim is a form of horizontal governance. Specifically, those stakeholders who are not party to the exchange may govern value chains through the mobilization of values-based standards (Loconto 2010). In this way governing power within value chains can extend horizontally as well as vertically.

As these four cases illustrate, the concept of drivenness is more complex than often claimed. A variety of contextual and embedded relationships have resulted in differing patterns of governance even within value chains. I summarize each of the four chains in turn. For the ETI value chains, the reliance on the traditional routes results in a highly captive, buyer-driven GVC. Participation or input by upstream actors is absent in most cases and textually mediated in those cases where their input is requested (i.e., via self-assessment questionnaires). However, producers are exhibiting power in some negotiations present in this chain, particularly with regards to their compliance with these standards and their ability to remain within the network despite non-compliance. The Tyhpoo QAP is a first party assurance scheme while the ETP has also taken the form of a corporate social responsibility exercise for its members. The lack of a consumer label has thus lessened the legitimacy of the ETP programme as an effective external check on suppliers, which may explain why many of its original members have either left the partnership or have also joined the Fairtrade and RA systems. The MOUs with RA, Fairtrade and UTZ exemplify the changes that ETP is making in its practices and perhaps represent an attempt to exert more power in a twin-driven value chain.

The fair trade value chain is the most complex of these four chains. Dolan (2008; 2010) has critiqued the Fairtrade tea value chain, based on its assumed "fairness" and focus on relational governance. She claims that "the reliance on the auction system in much of Fairtrade tea marketing militates against the longer-term buyer commitments and the symmetrical supply chain relations that relational governance implies" (Dolan 2008, 308). This study has found a relatively low percentage of Fairtrade sales in Tanzania (average 8% of certified production) and much of the certified product is sold through traditional marketing channels. This practice reduces fair trade governance to market-mediated relationships. However, the diversity of trading routes within fair trade systems illustrates the difficulty involved in relegating fair trade to a pure definition of governance. For example, we also see examples of hierarchical governance in the case of Herkulu and relational governance in the relationship between Cafèdirect and its producer partners.

The organic value chains exhibit both hierarchical and relational governance, dependant entirely on direct purchasing relationships and sharing of information between buyers and producers. Moreover, there is a paradox in the Tanzanian context between policy and practice. While much work has been completed to create an "enabling environment" for organic agriculture, this institutional context does not necessarily influence the organic tea value chains. Finally, we return to a classical case of buyer-drivenness with the RA standard where decisions made at the corporate level dictate which system to join and facilitate the certification of hierarchically governed value chains. Moreover, the "demand-driven" approach that RA used to roll out its certification, in addition to the intensive training required by farmers before meeting the certification requirements, inhibited the ability of producers to initiate involvement in this value chain and gain access to more lucrative direct sales contracts. However, the relational governance relied up by Taylors of Harrogate adds a new layer to the RA system and presents a case where alternative power relations between value chain actors might develop.

6. Conclusions

As explored above, there is no clear pattern in the relationship between standards regimes and governance relations. What are apparant are pragmatic strategies to influence power within the market through the inclusion of different actors into "value-added" chains. Moreover, only Organic requires certification and separation of certified and non-certified products throughout the chain beyond the packing stage into distribution; no standard covers the retail end of the chain. These shortcomings further illustrate why there is such variation and complexity in the governance of these chains, thus making the characterization of twin-drivenness over buyer-drivenness incomplete.

These conclusions suggest that despite claims about the ability to change the relationships amongst value chain actors through certification systems, most of the old networks remain. The certification systems appear to be adding additional buyers to GVCs that were already governed by highly relational mechanisms. These conclusions thus call into question some of the claims made by standards development organizations as to their abilities to change practices.

Notes

This chapter is based on work supported by the National Science Foundation under grant #SES-0924202 and an Institute of International Education US Student Fulbright Fellowship. Any opinions, findings, and conclusions or recommendations expressed in this material are the author's, and do not necessarily reflect the views of the National Science Foundation, the Institute of International Education or the US Department of State. Some names have been changed to protect the anonymity of research participants. The author thanks Dr. Lawrence Busch and Dr. Emmanuel Simbua for comments on early drafts.

1. During the time of data collection, three of these factories were not operating: Mwakalele, Chivanjee, and Lupembe. This was due respectively to insufficient leaf, rehabilitation, and legal disputes.
2. I differentiate between Fairtrade and fair trade throughout this chapter. Fairtrade refers to the standard and organisations affiliated with the Fairtrade Labelling Organizations International (FLO), while fair trade refers to the concept and other groups not associated with FLO.
3. The Fairtrade Tea standard and statistics cover black, green, white and oolong tea, they also include herbal infusions (i.e., camomile, hibiscus, mint), spices and rooibos tea.

4. Retroactive certification is defined as when a buyer has bought product from a certified producer or conveyor under ordinary conditions (non-certified), and wants to convert it into a certified product after the transaction.
5. Deutsche Gesellschaft für Technische Zusammenarbeit (GTZ) is the German Development Agency.

References

Baffes, J. (2003). *Tanzania's Tea Sector: Constraints and Challenges*. Washington, DC: World Bank.

Bair, J. (2009). *Frontiers of commodity chain research*. Stanford, CA: Stanford University Press.

Bedford, A. et al. (2002). *Managing social responsibility in tea and cocoa value chains*. London: Resource Centre for the Social Dimensions of Business Practice.

Blowfield, M. (2004). Ethical Supply Chains in the Cocoa, Coffee and Tea Industries. *Greener Management International* 43: 15-24.

Bombay Burmah (2010). *The Bombay Burmah Trading Corporation Limited—Fair Trade* 2010 [cited February 4 2010]. Available from http://www.bbtcl.com/fairtrade.html.

Conroy, M.E. (2007). *Branded! How the "Certification Revolution" Is Transforming Global Corporations*. Gabriola Island, BC: New Society Publishers.

Dixon, J. (1999). A cultural economy model for studying food systems. *Agriculture and Human Values* 16 (2):151-160.

Dolan, C.S. (2008). In the Mists of Development: Fairtrade in Kenyan Tea Fields. *Globalizations* 5 (2): 305-318.

—. (2010). Virtual moralities: The mainstreaming of Fairtrade in Kenyan tea fields. *Geoforum* 41 (1): 33-43.

ETP (2009). *The Ethical Tea Partnership Standard*. London: The Ethical Tea Partnership.

Id. (2009). Talking Tea-Issue 6. In *The Newsletter from the Ethical Tea Partnership for Tea Producers and other Stakeholders*. London: The Ethical Tea Partnership.

Id. (2010). *Enabling change across the tea sector*. London: Ethical Tea Partnership.

FAO (2009). Agricultural Sector Reforms in Tanzania: Perspectives from within. In *Project Report—"Strengthening National Capacity in Agricultural Trade and Trade Negotiations" (GCP/URT/056/IRE)*. Roma: Food and Agriculture Organization of the United Nations.

FLO (2009). *Generic Fairtrade Standards: Small Producers'—Current version: 15.08.2009*. Bonn, Germany: Fairtrade Labelling Organizations International.

Id. (2010). *About Fair Trade*. Fairtrade Labelling Organizations International (FLO) 2010 [cited January 23 2010]. Available from http://www.fairtrade.net/.

Id. (2010). *Fact Sheet: Fairtrade Standard for Tea*. Bonn, Germany: Fairtrade Labelling Organizations International.

Fold, N. and M.N. Larsen (eds.) (2008). *Globalization and Restructuring of African Commodity Flows*. Uppsala, Sweden: Nodiska Afrikainstitutet

Freidberg, S.E. (2004). *French beans and food scares: culture and commerce in an anxious age*. New York: Oxford University Press.

Friedland, W.H. (1984). Commodity systems analysis: An approach to the sociology of agriculture. In Schwarzweller, H.K. (ed.), *Research in Rural Sociology and Development*. Greenwich, CT: JAI Press.

Gereffi, G. et al. (2005). The governance of global value chains. *Review of International Political Economy* 12 (1): 78-104.

Gereffi, G. and M. Korzeniewicz (1994). *Commodity chains and global capitalism, Contributions in economics and economic history*. Westport, CT: Greenwood Press.

Gibbon, P. and S. Ponte (2005). *Trading down: Africa, value chains, and the global economy*. Philadelphia, PA: Temple University Press.

Guthman, J. (2009). Unveiling the Unveiling: Commodity Chains, Commodity Fetishism, and the "Value" of Voluntary, Ethical Food Labels. In Bair, J. (ed.), *Frontiers of Commodity Chain Research*, Standford, CA: Standford University Press.

IFOAM (2009). *The Principles of Organic Agriculture*. International Forum for Organic Agricultural Movements (IFOAM) 2010 [cited January 23 2009]. Available from http://www.ifoam.org/about_ifoam/principles/index.html.

Islam, M.S. (2008). From pond to plate: Towards a twin-driven commodity chain in Bangladesh shrimp aquaculture. *Food Policy* 33 (3): 209-223.

Loconto, Allison (2010). Sustainably Performed: Reconciling Global Value Chain Governance and Performativity. *Journal of Rural Social Science* 25 (3):193–225.

Loconto, A. and L. Busch (2010). Standards, techno-economic networks, and playing fields: Performing the global market economy. *Review of International Political Economy* 17 (3): 507-536.

Loconto, A.M., and E.F. Simbua (2012). Making Room for Smallholder Cooperatives in Tanzanian Tea Production: Can Fairtrade Do That? *Journal of Business Ethics*:1-15. doi:10.1007/s10551-011-1101-9.

Neilson, J. and B. Pritchard (2009). *Value Chain Struggles: Institutions and Governance in the Plantation Districts of South India.* West Sussex, UK: Wiley-Blackwell.

Ochieng, C.M.O. (2008). Comparative capitalism and sustainable development: Stakeholder capitalism and co-management in the Kenyan fisheries sub sector. *Natural Resources Forum* 32: 64-76.

Pirsch, J. et al. (2007). A Framework for Understanding Corporate Social Responsibility Programs as a Continuum: An Exploratory Study. *Journal of Business Ethics* V70 (2): 125-140.

Ponte, S. and P. Gibbon (2005). Quality Standards, Conventions and the Governance of Global Value Chains. *Economy and Society* 34 (1):1-31.

RA (2010). *Our Values and Vision for the Future*. Rainforest Alliance (RA) 2010 [cited January 23 2010]. Available from http://www.rainforest-alliance.org/about.cfm?id=values_vision.

Id. (2010). *Rainforest Alliance Certified Seal to Appear on Tetley Tea* 2010 [cited February 17 2010]. Available from http://www.tetley.co.uk/Why-tea-is-great/All-about-tea/Rainforest-Alliance

Simbua, E.F. (2006). The role of the Smallholder Tea Based Production Systems in the Tanzanian Tea Value Chain. In *Annual Meeting and Scientific Conference of the Agricultural Economists Society of Tanzania (AGREST)*. 14-17 August 2006, Morogoro, Tanzania.

Tallontire, A. (2007). CSR and regulation: towards a framework for understanding private standards initiatives in the agri-food chain. *Third World Quarterly* 28 (4): 775-791.

Tallontire, A. et al. (2011) Beyond the vertical? Using value chains and governance as a framework to analyse private standards initiatives in agri-food chains. *Agriculture and Human Values* 28 (3): 427-441.

TBT (2009) *Performance of the Tea Industry in Tanzania*. Dar es Salaam, TZ: Tea Board of Tanzania.

TransFair (2010). Jahresbericht 2009–Ausblick 2010 (Annual Report 2009—Outlook 2010). Köln, Germany: TransFair e.V. / GoodWeave.

Typhoo (2010). *Ethical Policy—Making Good Tea*. Available from http://www.typhootea.com/ethical-policy.html.

Unilever (2007). *Unilever commits to sourcing all its tea from sustainable ethical sources* 2007 [cited May 25 2007]. Available from

http://www.unilever.com/mediacentre/pressreleases/2007/sustainable-tea-sourcing.aspx.
van Reenen, M. et al. (2010), *Tea Barometer* 2010. Leiden, NL: Tropical Commodity Coalition for Sustainable Tea, Coffee Cocoa.

CHAPTER FOUR

A CONSTRUCTIVIST VIEW ON THE QUALITY OF FOOD: ARGENTINEAN EXPERIENCES

MARCELO CHAMPREDONDE AND JOSÉ MUCHNIK

Abstract

Quality of regional food needs to be approached from diverse points of view that integrate the complexity of its multiple dimensions. So far, there have been two main contrasting approaches: the objectivist and subjectivist schools. The objectivist school points toward the materialization of quality and its measurement through a series of objectifiable variables. The subjective school, unilaterally at times, focuses on the role of symbolic values associated with the production, elaboration and consumption of food.

The present approach aims to consider quality of food from a dual constructivist perspective. On one hand, it aims to understand quality as a social construction, i.e., "a process that proposes not only the products but also the interactions occurred in the definition and control of said products" (Casabianca and Valseschini 1995). On the other hand, it approaches the process of qualification of the product from a scientific and constructivist view: reality (in this case, the quality of food) cannot be known as what it is. This is due to the fact that when an individual encounters an object, he/she cannot but notice and order the data the product offers into his or her cognitive framework.

Three field experiences, Argentine Pampas beef, Salami of Colonia Caroya and Amblayo cheese are presented.

1. Introduction

Within the present context, the location and dislocation of production processes presents a great challenge from the economic and social view as well as from the cultural view. This challenge is particularly acute in the agro-food sector since the quality of food is directly affected. A wide range of questions arise as a consequence of this evolution, ranging from the health of the consumer to the gustatory evolution of products. The main goal of this work is to discuss the mechanisms of territorialisation or deterritorialisation of the quality of food. This approach focuses on objectifiable variables (physical, biochemical, rheological and other characteristics) and subjective variables (representations, values, tastes…). Both types of variable depend on the perception of subjects, i.e. different types of consumers.

The social change of the 20th Century led to an evolution in the conceptualization of quality of agro-food products. At the beginning, quality, related to the industrialisation of agriculture and the agro-food industry, was associated with the line production of homogeneous and safe goods. With the transformation of modern and post modern societies, the concept of food quality developed to include aspects such as the re-evaluation of traditions, cultures and local know-how, the search for a greater environmental care and consumption of "healthy" foods.

Up to today, views on quality have been divided into two major schools. The first school could be qualified as objectivist. It perceives the materialisation of quality and its measures by means of a series of objectifiable variables, for example, the utilization of biochemical and molecular tools (functional proteomics and genomics) to study the physiological mechanisms for quality improvement of raw materials. The second school, subjectivist, focuses on the role of the symbolic values associated with production, processing and consumption of food. A great deal of social science research is focused on the place of food in social relations and the differentiation amongst individuals.

The choice of a constructivist perspective to analyze the role of territory in the quality of food is based on the fact that this quality is the result of the interactions between different subjective and objectifiable variables. As Lévi-Strauss remarked, "Food must be not only good to eat, but also good to think". So the quality of agro-food products has to be understood from a point of view that integrates the complexities of its multiple dimensions.

For the development of this analysis, three contrasting Argentinean experiences[1] are evaluated: (i) Beef from the Pampean region, (ii) Salami from Colonia Caroya, South of the province of Córdoba, and (iii) cheese of Amblayo from Calchaquies valleys in the province of Salta. The first product is acknowledged and consumed national and internationally. The second product belongs to a mainly regional and tourist market. The third product is consumed mainly regionally. Using these three cases we make a contrastive analysis of the role of the objectifiable and subjective variables that interact in the process of the construction of quality.

This chapter will first discuss views and theories of quality. It will then analyse the three cases, taking into account the historical evolution and the role of objectifiable and subjective variables. The comparison of the technical and social trajectories will finally reach conclusions and will discuss the plausibility of a constructivist view on quality.

2. Quality and qualification: views and theories

Views on quality and qualification of foods have progressively incorporated the human being and his or her perceptions. From a strictly mechanical view, views have evolved towards including interactions between objects and persons. Consumer perceptions have been added to the measurable particularities of the products. In this perspective, Casabianca and Valeschini (1995, 12) proposed the concept of "*bipolar representation of quality*". They identified a pole of objectifiable characteristics and another "*pole of subjective variables marked by the consumer*". From a constructivist view, not only is the existence of both quality poles acknowledged, but also its character of being constructed by humans. In other words, if there is an objectifiable dimension or "pole", it is not free of human subjectivity. The other fact that is acknowledged is that both quality and qualification of food constitute situated constructions.

2.1 "Objectifiable" schools

The objectifiable variables are the characteristics of food that belong to sciences like physics, chemistry and biology, and to disciplines like genetics and microbiology. All these are supported by mathematics, with the purpose of establishing the orders of importance and the frequency or probability of occurrence of the events. Examples are the concentration of casein in milk, the tension of water in the salami or the colour of the rib eye, the microbiological populations related to food safety, the rheological (texture) characteristics of a dough, and the set of sensory characteristics. So, from an objectifiable point of view, the sensory characteristics of food are formalised by measurements using tools (See Cizalla Warner-Bratzler for measuring cut resistance of meats, and the "electronic nose" for the detection of food aromas), and by panels of trained tasters. Conclusions are based on statistical elaborations of "objectified" information.

Recently, there has been notable progress in the objectivation of new variables, such as the detection of alergogenic substances, or the presence of chemical components with positive or noxious (contaminant) impact on human health.[2]

The "objectifiable" views are supported by logics of "standardisation": the standardisation of the "living beings" (Casabianca and Letablier 1995), of the relationships among actors (Thévenot 1995), and techniques (Dassa and Marillard 1995). "Best practice" is set out in manuals that lay down, for example, materials for use in tools, equipment for use by workers, food safety practices, optimum method of a slaughter for meat tenderness (organoleptic quality), animal diet and how it impacts on the profile of the fatty acids of meat and its potential impact on human health, the quality of the animal protein (nutritional quality) and health and safety precautions for manufacture.

From an objectifiable point of view, based on a "positivist view of science", "quality can be defined as the "degree to which a set of inherent characteristics fulfils the requirements" (Figuié and Bricas 2002). This definition belongs to concepts of quality that underlie ISO standards (International Organization for Standardization). The ISO standards were created with the purpose of inspiring confidence in users, based on the homogeneous and predictable characteristics of the products or services certified. This goal is achieved by the standardisation of the processes and/or the products. Many contributions since the 1960 have used this concept. Here, quality is defined in terms of "fulfilment of requirements". The proposed goal is to reach a level of "zero error" in the management of quality of products.[3]

Again in the objectifiable view, the concept of quality is related to "superiority or excellence".[4] Schools that propose the implementation of management for total quality are based on "the use of different means in order to achieve a desired level of quality" (Juran and Gryna 1994). The stages necessary to achieve this goal are: a) the setting of goals, b) planning of the organisation and the means, c) motivation of personnel, d) start-up, e) evaluation of results, f) analysis of deviations, and g) corrective actions.

Thus, the definition of quality corresponds, on one hand, to a vision of standardisation of the processes and products, in which enterprises' competitiveness is based on the assurance of market niches and on the lowering of production costs.[5] On the other hand, innovation in quality derives from better new product designs than improve on competitors' or that reduce production costs. This type of innovation is supported by the skills of qualified professionals, often in networks searching for solutions to society-level problems.

The objectifiable conceptions of quality, based on the positivist views, intend to formalise and standardise the processes and knowledge involved. However, even in this view, subjective variables are not entirelyabsent. An important constraint in the characterisation of "objectifiable" variables of products is the implicit subjectivity in the stages of decision of sampling techniques and the measurement techniques employed. There is also human error, which occurs in any observation technique. So the term "objectifiable" instead of "objective" is purposely used in this chapter to highlight the influence of the objectivation process.

2.2. "Subjectivating" schools and "modern times"

Subjectivating views acknowledge the importance of the symbolic dimension of food for the consumer. Foods are acknowledged by their function as I identitary reference; they contribute to the individualisation of persons, to distinguish them from peers, and at the same time to define their belonging to a given social group.

Marcel Mauss (1935) characterised nourishment as a "complete social fact", a fact that reflects the changes undergone by the complete social set. Claude Lévi-Strauss (1962) revealed the significance of nourishment in the belonging to a culture and a territory. This manifests through representations and rules of behaviour structured around food, regulating what can and cannot be eaten and culinary practices, and dictating rules for eating, the place of the food and the nourishment in the organisation of the social life. More recently, authors like Claude Fischler (1996) highlighted the specificity of foods as regards other consumption goods, since it is the only good that is "in-taken" (introduced into the body), which gives it a specific role in the construction of the identity of individuals and societies.

The profound changes in the modern societies have accentuated the importance of the subjective perception of the quality of food. This tendency, strongly accentuated from the 1960 onwards, has led actors of the different agri-food systems to consider the new symbolic components in the production/processing processes of their products as well as in their place in the markets. On one hand, growing urbanisation and social transformation are accompanied by a new conception of food. The modern urban consumer imposes "imperious demands as regards the purity and constant quality of products. He wants white bread and dairy products whose flavour is always the same... and is thus orientated towards standard production" (M. Sorre 1943-1952, quoted by Delfosse 2006, 4).

On the other hand, the internationalisation of markets requires that "products for export are not to agree to the local or national taste, but should be neutral, corresponding to the tastes of consumers of any country, like for example Danish butter" (ibid., 4).[6]

The nature and the complexity of these processes and the consequent transformation in the symbolic quality of food are particularly clear in beef, for example. These transformations have been researched by authors such as Poulain, who writes that during the first decades of the post-war period, once access to the food was secure, the consumption of meat was seen as access to wellbeing. "Meat, the par excellence food, is loaded with a strong social symbolism: eating it on a daily basis is to access wellbeing, is to get revenge" (1997, 106). At that stage of modern history, *boeuf*[7] and veal were perceived as noble meats. The consumption of "*veau de lait*" (suckling veal) was associated with the aristocracy.

Later, along with growing "industrialisation" and "artificialisation", in the 1970, consumer associations demanding an end to "hormones in meat" grew up. The 1990s saw food scares such as dioxin in chicken meat and "mad cow disease" (*Bovine Spongiform Encephalopathy*). There was an increase in relatively negative social representation of industrially produced food, seen as potentially noxious to health. New discourses and symbolic values emerged.

Two of these new discourses, often associated, opposed the image of industrial food. There was the start of a search for "natural"[8] food from "ecological" and "organic" environment-friendly farming. At the same time, vegetarian, naturist and other diets became more widespread. In opposition to the tendency to a loss of identity of food, there was a search for foods anchored in the "territory", i.e., in local traditions, or the craft preparations, based on supposed family traditions. This process can be seen in the slogan "Grandma's recipe".[9] Terms like "typical", "regional", "local", "homemade", "natural", "craft", "of origin" and others constitute expression of new and diverse ways of qualifying products.

Other aspects of the "artificialisation" of agriculture are related to its impact on lifestyle and landscape in rural areas and the loss of local know-how. These phenomena are also incorporated as subjective components of the quality of food (Hirczak 2007).

The changes in the symbolic value of food and nourishment were also reflected in models of beauty. Until the first half of the 20th Century, the opulence of flesh, as in the paintings of Renoir, prevailed as an indicator of high social level. Then, comes the Modigliani model with thin and stylized figures. Obesity and fatness have now become an indicator of health problems and low social level. Poulain writes "A rationalist diet, medicalised, is vulgarised and announces the divorce between what is good and what is healthy, between gastronomy and dietetics" (Poulain 1997, 106).

These words also reveal the growing importance of another dimension of quality; the 'medical' function [10] of food. Today, food is consumed not only to prevent disease, for example, foods containing certain vitamins, minerals, antioxidants, etc., to prevent cancer, where there is overlap between nutritional and therapeutic quality. It is also consumed to cure illness (for example, garlic to lower blood pressure), as a stimulant (coffee, mate, or tea), as a relaxant (linden tea) or an erotic enhancer; ginger (*zingiber officinale*) or muña muña (*Satureja parvifolia*) in Argentina and Bolivia.

Because of the important social changes, social sciences have become more important in the characterisation of food quality. All symbolic elements associated with an area, culture or time can now be included among the variables that determine the subjective quality of food. They include transformed know-how, the physical and human context of processing, tools, ingredients, festivities, historical tales, etc. These symbolic elements are present not only in the imagination of the consumer but also for the set of actors interacting in agri-food systems.

Along with these new ways of qualification of food, tools and facilities in which these processes are carried out acquire symbolic value. For example, in the elaboration of Amblayo cheese the belt-like girth (*cinchón*)[11] in which the cheese is pressed is considered as the traditional mold. In Colonia Caroya, the traditional environment for curing the salami is the basement.

2.3 Constructivist view: quality as an integrative variable, the qualification processes

One element appears central in all schools and views; quality implies an inevitable process of evaluation i.e., quality of a product is the result of a process, defined in terms of qualification. In every process of qualification, the human component given its interactive character plays a central role:

> as with product identity, quality is not intrinsic. Quality is defined as "in relation to". It is the act of qualification that reveals the quality of the product by means of the interactions between the product and the criteria used for judging it. It depends on the criteria on which it is based, how it is chosen, who judges it… (Muchnik et al. 2005)

For researchers and agents of intervention, being framed in a constructivist view implies participating "actively in the construction of reality, giving it a sense" (Gundolf 2006, 101). From this point of view, product qualification exceeds mere product description in a list of characteristics or requirements to be fulfilled. Product quality is thus the result of a process of interactions between the product and the human beings that participate in its qualification.

This reflection is based on two interpretations complementary to the constructivist term. (i) On one hand, considering quality as a social construction, i.e., "a process that places not only the products but also the interactions around the definition and control of those products" (Casabianca and Valseschini 1995). These interactions regard the set of actors: extractors/producers, transformers, production workers, distributors and consumers. (ii) On the other hand, the process of qualification[12] is seen from a constructivist scientific (as opposed to the cognitivist) point of view. This postulates that reality (food quality) cannot be known as what it is. This is because each individual approaching an object through his or her cognitive framework perceives and orders the data of the object for themselves.

Under the premise that there is no quality without qualificator, quality is considered as the result of a human construction. From a constructivist point of view (Manheim 1936, Schutz 1967, Berger and Luckmann 1967), social objects do not exist as a given element "but are built, negotiated, reformed, adapted and organised by human beings in their effort to give sense to what is produced in the world". For the constructivism there is no exact description of how things are, because reality (in this case, quality) does not exist without the subject.

From this constructivist perspective, the process of construction of quality is considered as part of a learning process, both individual and collective, in which the representation of objects is built from the knowledge of the actors and the symbolic values they attribute to the qualified object. So, the diverse actors develop their actions in the context of a *located cognition*.[13] F. Laville postulates the hypothesis that the rationality of every act derives "from the interaction of an agent with his environment (located action) and the interactions between members of a group (distributed cognition)" (Laville 2000, 1328). This theoretical contribution helps to interpret the particularities of the construction processes of quality and the qualification of products, characterised by the presence of multiple uncertainties.

Faced with these uncertainties, it is explicit that "in a given situation, all the faculties distinguished by the cognitive psychology (perception, language, action, reason, memory) act differently" (ibid., 1311). That is, the qualification of a product may vary in time, according to the context in which it develops, and the perception of the actor in that situation. So at every qualification, actors do not perceive object properties, but "the possibilities the objects offer them" (ibid., 1312). There are thus cognitive limitations to be considered, as regards the actors involved in the process of qualification of food, and in the context in which this process takes place. Interactions make possible the estimation of quality of a product, implying both the object and the qualifier, the qualifier's perception supported by a specific criteria.

Another component to be considered in the process of quality construction and qualification is what Laville calls *distributed cognition*, i.e., the cognitive resources of the environment (physical, social, linguistic, cultural and technological), and the social distribution of the cognition. For this, the type of distribution of the knowledge orientated towards the qualification of supplies and the "final product" has to be considered during the different stages of processing.

This knowledge oriented towards the construction of quality is coordinated among the actors involved in the different stages of an agri-food system by means of phenomenon called translation. This is the development of a

> language specific to the qualification of the product... that contributes to create equivalences between the consumers' expectations and the elements involved in the definition of quality in the previous stages of the agri-alimentary system (Champredonde 2001, 196).

This translation "is framed in the process of specialisation (tasks division), in processes of recollection/extraction/production, transport, transformation/preparation, distribution and consumption of food". "This translation adopts the specific terms of each technical stage" (ibid., 196). So, for example, the dimensions used to describe the quality of a steak are not the same as those used to describe the quality of a half carcass or a head of cattle.

The particular quality of a territorial product is therefore result of multiple interactions between man and matter in a given natural context (Muchnik 2005). These interactions occur a) in the coordination of knowledge distributed among the actors that compose an agri-food system (translation), b) in the interaction between these practices and the material components, and c) in the interaction of knowledge and supplies with the local environment (human and environmental). In the case of territorial food, consumer knowledge is the guiding aspect acting through consumer approval or disapproval of the qualities offered by the rest of the actors.

3. Qualities throughout territorial transformations

The three experiences are analysed considering: (i) the historical process of rise in production activity, (ii) the construction of quality, and (iii) the evolution pathways.

3.1 Pampas beef

Rise in production activity

The history of meat is the history of the Pampa itself. Until the 15th Century, the people that inhabited the Pampas flatlands based their diet on the meat of native species. With the arrival of the Europeans, these species were partially displaced by introduced species, which competed for the exploitation of the fodder resources, and which modified the native cultures.

By the middle of the 19th Century, the European Industrial Revolution had a strong impact on the region, generating an immigration current that had a determining influence on the culture and economy of the territory. This process propelled the development of sheep and the bovine cattle and the moving of the frontier. In 1879, the Conquest of the "Desert" put an end to the millenary presence of native cultures in this region. In this new stage, the *gauchos* became the farm workers taking care of cattle on the big ranches. The culture of the *gaucho* blended into the different immigrant cultures expelled by European industrial development dreaming of a better life in South America.

Construction of quality

A re-examination of history shows that Argentine Pampas meat international markets centers round the factors that contributed to its particular quality. Orientated first to the United Kingdom, the meats produced at the beginning of the 20th Century were characterised by the predominance of Scottish breeds (locally known as British) such as Shorthorn, Hereford and Aberdeen Angus. The biotypes selected in this region were characterised until the middle of the 1970s by the predominance of a small cattle with a big capacity for fattening. This encouraged adaptation to the pasture systems and the production of tender marbled meats under such conditions.

Adjoining this "objectifiable process", knowledge related to the diverse activities in the agri-food system and a specific food culture developed. At the early stages, knowledge factors concerned selecting biotypes adequate for size of cuts and fat depth at young age (associated with tenderness and juiciness of meat), the castration of male calves at the age of 8 months (for tenderness), the procurement of pasture resources adequate for growth and fat depth at young ages (for tenderness) and the visual estimation of the fat depth adequate for slaughtering.

In retailing live cattle, the consignataries[14] developed knowledge referring to the estimation of live cattle quality. This implies estimating the degree of fattening, the weight of the half carcass and the hanging requirements. In the cutting stage, slaughtering house employees and butchers acquire the knowledge that allows them to obtain typical Pampas cuts, for good presentation. This has to do with the capacity of local consumers to select good meat. For example, when buying a typical Pampas cut known as *asado* (short and spare ribs), the buyer observes the width of the rib. A very wide rib is associated with an older animal and therefore a lack of tenderness. The colour of the meat is another aspect associated with tenderness. In the Pampas region, the consumer associates a pinkish to light red colour with an early age of slaughter. Another aspect evaluated in the buying is the degree (excess or shortage) of fattening.

Other types of knowledge are set in motion when preparing the barbecue. Cooking times need to suit different tastes. Some parts have to be juicy (but not semi raw) and others have to be "*a little dry*" (well done). The meat is cooked by rotating it only once to complete its cooking on both sides. The cut called *asado* is first cooked on the side of the bone and then on the side of the meat.

Evolution pathways: from the gaucho to the mixer[15]

The case of Pampas meats represents the evolution of a product known for high quality on the international market. In the objective components of the quality, tenderness, juice and flavour are its highlights. But the most important component of quality, on the international market is the association with the mythical Pampa, an immense flatland covered in pastureland and travelled by the no less mythical *gaucho* on horse. On the internal market, mainly composed of Pampas consumers, bovine meats are associated with the extensive pasture systems as a natural and healthy product. But modernity has partially negatively modified the symbolic quality of meat; consumption is now linked to high of cholesterol and uric acid levels.

If at a subjective quality level some changes can be observed, the most important transformations have been recorded at an objectifiable quality level. Diverse territorial changes contributed to the transformation of the objectifiable quality of Pampas meats. On one hand, the progressive process of agriculturisation has contributed to a movement of bovine farming to less productive areas of the Pampas. There have also been transformations imposed by modernity and hyper-modernity (Champredonde and Abaladejo 2008, 9).

Bovine farming has seen an increasing use of nutritional supplements in fodder, mainly silage of corn, sorghum and barley. Pellets of sunflower skins and soy flour are also used. The direct impact of this process is the increase in the probability of obtaining tender meats with a more marbling. The multiplication of intensive production systems has led to cattle more homogenous as regards conformation and particularly depth of fattening.

This results in a better presentation of the meat to the eye of the local consumer; it presents light red colouring, small size of the cuts, and in the *asado*, short ribs. This tendency can become a real threat to the meats of pasture systems, as the consumers get used to extremely tender meats, and reject meats that are less tender. As a negative aspect, this type of meat is more prone to develop alterations of scent and taste (liver taste, pork taste, etc.) and is also prone to a higher instability in oxidative processes (stale taste), specially in processes of maturing superior to 28 days. Other research carried out by INTA (Instituto Nacional de Tecnología Agropecuaria) shows a negative impact on the nutritional quality of meats. The increase of saturated fatty acids of meats originating from intensive systems enhances the risk of cholesterol increase in the consumers. The difficulty with this "non quality" is that it is hardly detected by the consumer during consumption.

In this case the most important changes are occurring in the objectifiable level of meat quality. In Europe events have taken a different turn. But in spite of the great changes in objectifiable quality, the image of Pampas meats as a natural product and the anchoring in the habits of the local consumers mean that the process is currently having low impact on its symbolic quality.

3.2 Salami of Caroya

Rise in production activity

In 1878, approximately 60 families arrived from Friuli (Northwest Italy), and inhabited the Jesuit farming estate of Caroya. In the early years, they worked mainly on the systematisation of the plots destined to each family, and developed a crop irrigation system.

Today, the colony has 16.000 hectares divided into holdings ranging between 2 and 24 hectares. The local conditions allowed the cultivation of crops such as vines, fruit trees as well as vegetables such as sweet potatoes, garlic and potatoes in an extensive system. Over time, an urban fabric grew up in the middle of the colony, characterised by the main avenue and its parallel rows of centuries-old plane trees.

Construction of quality

As they were well known to the Friulian immigrants, sausages have been made since the beginning. Over time, a salami of which the recipe varies slightly from family to family was generalised among the Caroyan families. It is made of bovine and pork meat, bacon, sweet white wine, ground pepper, salt, potassium nitrate and garlic soaked in wine.

Since the foundation of the colony, the salami has been made at home by farming families. Farms breed pork fed on alfalfa and corn, and slaughter a cow for the occasion. The slaughter (*carneada*) is a family and social event. The making and salting of the sausages is carried out in the winter months, and generally takes place in private garages. The sausages are destined for local consumption and in some cases for sale in nearby areas. The volume of salami produced annually is less than 800 kg. All salami is matured in a basement.

The local knowledge that allows the production of a good Caroya salami is distributed between knowing how to chose the supplies, specially the meats, preparation of the mixture, stuffing (generally done by men) and tying (generally done by women), and curing. Curing has the most impact on the quality of the salami, so humidity and room temperature is determinant. Most people responsible for this stage (mainly men) know whether to increase or diminish the humidity and temperature by touching the external surface of the salami.

In the traditional systems, the regulation of temperature is achieved by opening and closing the basement windows at different times of day. The humidity is modified by using hot coals when there is need for a drier environment. The exact temperature (between 15 and 18 Celsius degrees) and room humidity (over 70 percent) allow the normal development of two parallel processes: the fermentation of the mixture through the presence of bacteria and the progressive loss of humidity.

An important component in the maturity of the salami is the development of a mould in varied shades of grey, known as *mufa blanca* (white mould). This mould is an indicator of the curing process and contributes to the particular taste.

Evolution pathways: from the basement to the cold store

The change from household production for internal consumption to commercial production took place in the 1950 because of a local economical crisis caused by the loss of a fruit harvest as a result of a hailstorm. This crisis encouraged some families to switch from salami for internal consumption to bigger scale commercial production. Over time, output rose in terms of the number of producers and volumes each produced. There was in fact great variation in the scale of production and equipments and manufacturing processes.

The increase in output for commercial sale was mainly led by the local butchers, who often supplemented their income through making salami and *bondiolita*. In the last decade, many of were transformed into big producers specialising in salami, and also supplying products such as *bondiolita*, *jamoncito* and codeguin, etc. These enterprises produce salami year-round in volumes that vary from 1.800 to 150.000 kg annually. The labour is predominantly family members, and in the big enterprises is mainly hired. In most of these cases the distribution of labour is similar to that in household production.

However, the substantial increase in the volume, in industrial units, has led to a lower capacity in selecting meat according to quality, in separating the nerves on each processed cut and manage the curing. The central element of the original recipe has seen some variations. These affect, first, the proportion of bacon; to satisfy consumers worried by the impact of saturated fats on health, the percentage of bacon has decreased from 30% to 15 to 25% of total weight. The proportions of beef and pork vary according to traditional family recipes and variations in seasonal price and curing duration desired by the producer. Pork makes for a longer drying time than beef.

The changes which are less approved by the local population are, first, the use of synthetic gut instead of natural gut for casing. This technical resource is relatively little used and is confined to big volumes. The second aspect constitutes a taboo subject for the local population. The use of additives and ingredients to speed up fermentation (bacteria starters) or to imitate the consistency of a mature salami in less time (gluten) is very much disapproved of at local level. Among producers, "It is not talked about". Among consumers it stirs ghosts. But the characteristics of the product often make it appear that additives are being made use of. In some cases they result in the loss of organoleptic quality in the salami.

As regards facilities, the garages used slaughtering in the past have often been converted into processing rooms. To comply with current legislation, today they use tiled rooms and they keep processing rooms separate from others. Tools have mainly evolved in the type of material they are made of: the current spectrum of tools varies from epoxy painted machinery and wooden tables to the legally approved stainless steel.

One of the major changes is that salami is now made the whole year round. The need to generate income the whole year round brought the technical challenge of how to make it continuously. Curing now has to take place in extremely variable conditions of temperature and humidity. Equipment is now needed for air-cooling in the basement.

Traditionally curing took place in the basement, but in recent years the basement has often been replaced[16] by industrial cold stores. At a symbolic level, the cold store is a synonym for industrial quality. Some artisan producers even say that the modification of the environment has a potential impact on the taste of the salami[17]. The duration of curing is another aspect that has been modified. In numerous cases, the "traditional" minimum duration (at least 18 days before consumption) is not respected). Salami is sometimes sold locally or shipped when it has been cured for barely over a week.

Social changes such as smaller family size and the desire for healthier diets has also had an impact on the presentation of the salami. On one hand, the proportion of bacon used has decreased too. This is because consumers prefer to buy smaller units and avoid keeping pieces of salami in the fridge for a long time (more than a week). On the other hand, a smaller size, especially diameter, accelerates drying. This factor combined with a higher proportion of beef in the mixture is leading to faster production times.

Other changes at consumer level have played an important role in the evolution of Colonia Caroya salami. Today, the Caroyans are tending to eat salami less mature. Forty years ago the majority would consume salami cured for 25 to 30 days but now a time-span of 18 to 20 days is more frequent. There has been an exponential increase in external consumers, people who travel Route 9 through Colonia Caroya or who "invade" the mountain zone adjacent to the city of Caroya. New selling points have in fact been opened on Route 9. Other external consumers are supplied with salami in their own cities throughout the country thanks to distribution by three retail companies. The most direct impact of this phenomenon is the lack of specific tasting skills among consumers, which means that different levels of quality can be sold. This has a repercussion in the quality offered by some producers. Some Caroyans actually say, "*Some people are producing inedible salami*". According to local people, there is a difference between homemade and commercial salami. Some Caroyans are even proud to say they have never eaten a commercial salami. Homemade salami is what is produced for local consumption in the farming area.

Another aspect of symbolic quality associated with volume is how the salami is tied. In large scale production it is difficult to respect the traditional three or four knots of a single string, which traditionally gives the salami its typical appearance.

3.3 Amblayo Cheese

Rise in production activity

The mountain range of Amblayo is situated in the province of Salta and is part of the called "Calchaquies Valleys". This is a semiarid region with strongly marked seasons. The main activity is goat farming and goat's milk cheese. This zone is located 140 km southwest of the city of Salta and three hours away, 2000 meters above sea level. The rural population is thinly spread and there are small centres of up to 150 inhabitants.

Amblayo producers would no doubt agree that a typical cheese is made of goats' milk, using natural rennet, and matured in molds of *cinchón* for at least 15 to 20 days. The area of production would embrace the pastureland occupied by the community of Amblayo and its area of influence.

Construction of quality

In the organization of labour within the family units, the production of cheese is mainly in the hands of the women. Because of the location and type of labour, the cheese could be catalogued as a kitchen product[18], which implies that knowledge is transmitted mainly from mother to daughter.

According to the interviews with producers, a combination of knowledge appears to take place, mainly on the occasion of marriage. The bride tends to adapt to the practices of her husband's family, but she brings an imprint of the acquired knowledge in her family. But the main source of the diversity in practice in maintaining the herd and producing cheese lies in the variations in humidity and particularly temperature registered between the different areas. Producers say that a cooler temperature and higher altitude facilitates cheese ripening. So managing the ripening process is complex and requires greater technical resources in lower-lying areas.

Making and ripening also need to be adapted to the composition of food intake and composition of crops, which reflect the seasons and environmental conditions. Basically, the best cheese is obtained from March onwards thanks to the better quality of the pasture.

Evolution pathways: from traditional mold to PVC

The major changes in the quality of Amblayo cheese are the result of local changes. First, there was the opening of a road in the 1960. Up to that point, the cheese was transported by mile to the railroad station, and from there by train up to the city of Salta. This change impacted on the main characteristic of the cheese. Second, the wheels of cheese weighed up to 8 to 10 kg to facilitate its transportation on the mule's back. This cheese was matured up to 12 months. It is around this type of cheese that the reputation of the cheese of Amblayo was built on.

Transportation by vehicle (refrigerated in the best of cases) led to the commercialisation of smaller wheels of less ripened cheese. Nowadays, the cheese tends to weigh from 1 to 1.5 kg has undergone from 5 to 15 days of ripening. Other territorial and technical changes had an impact. On one hand, there has been a change in the composition of the herd due to an increase in the breeding rates of sheep. This is because sheep's milk gives more cheese by litre and because sheep are more docile than goats. In addition, the physical capacity of the local population is lower because of the exodus of young people to towns. Sheeps' milk makes the cheese creamier and less strong-tasting.

At the level of production practices, technical advice from external specialists has impacted on the genetic composition of herds. Few producers today keep the "pure" genetic biotype. Also, new water management systems for irrigation and domestic use have generated changes in seasonality and the quality of fodder. The impact of these two technical changes on cheese quality has not yet been studied. Another of the technical changes affecting cheese quality is the growing use of chemical rennet. As regards the current use of metallic or PVC moulds, these allow the whey to drain away more slowly, which directly affects the texture, taste, and appearance of the cheese, particularly the marks on the side of the wheel. This is mainly associated with the subjective variables of quality.

As regards food safety, it does not appear to be a central concern for producers. Skill of milkers and cheese-makers along with conditions of low humidity and low atmospheric pollution are probably sufficient to neutralise the dangers posed by poverty of tools and facilities (adobe kitchens etc.). Research by the INTA (Chavez M. et al. 2009) show satisfactory parameters for health and hygiene quality of milk and production systems. Similar results would be expected for larger units if the same families were making the cheese.

Despite the big impact of local changes, the image of Amblayo cheese has not altered among the consumers and visitors in Salta. Regardless of the big changes in qualities, all output is easily sold in Salta. In fact, the municipal market of San Miguel in Salta is one of the other places in which the artisan product is commercialised, and where Amblayo cheese is object of imitation. This implies that there is little specific knowledge of the product by most consumers, especially the inhabitants of Salta. Only some consumers are capable of differentiating the typical Amblayo cheese from its imitations or some of the cheese produced nowadays in Amblayo that differs from the traditional quality.

A somewhat unhelpful element in the process of definition of a typical product is the cheese contest that takes place in the city of Isonza near Amblayo. The fact that its jury is composed of political and other personalities with little expertise in cheese tasting can lead to widespread misperceptions about cheese quality. It has been known for mixed milk cheese rather than a pure goats cheese to win the best Amblayo cheese award.

4. From the territorial food to the (re)construction of territorial qualities

The three cases help us to understand the multiplicity of factors that interact in the processes of territorialisation/deterritorialisation of the quality of food and offer indications for possible intervention. The understanding of the dynamics of the objectifiable and subjective variables and the interactions among them lie at the centre of the analysis. This type of analysis can clarify the pertinence of the intervention and the type of strategy to be implemented for re-territorialisation of quality, and can also give some idea of the probabilities of success.

4.1 Territorial quality is vanishing?

The three cases analysed show that there is a general tendency towards the relative loss of territorial food quality. In the cases of Pampas bovine meat and Amblayo cheese, this tendency involves a large part of production, and in the case of Caroya salami it involves half or more of the objectifiable and subjective components of quality. These changes are summarised in Table 4-1.

Table 4-1 Evolution of the two variables of quality for the three products analysed.

	Objectifiable Quality	Subjective Quality
Pampas Beef	Earlier ages and lower slaughtering weight. In steaks: strip loin steak, smaller rib eye, lighter colours. Increase of marbled and proportion of saturated fatty acid. Greater probability of tender meats.	Extensive systems. "Natural" foods. Healthy food (protein and iron) with negative impact on cholesterol. Negative impact of feedlots on the environment but without translation on the qualification of the meat.
Colonia Caroya Salami	Part of the curing in cold storage. Shorter curing times. Lower fat content. Smaller salami size. Partial use of new additives, with changes in flavour. Implementation of Good Manufacturing Practices in some producer enterprises.	Reputation maintained. Increase in use of terms "commercial salami" or "industrial salami" among local population. New tying on some of the salami.
Amblayo Cheese	Reduction in the size and the duration of ripening. Chemical rennet. Metallic or PVC molds of. Mixture with sheep milk. Changes in texture and taste. Problem of safety from shorter ripening?	Reputation maintained in the market of Salta. Mark from mold no longer appears on much production.

It can be observed first that the most important changes in the territorial quality of the products occurred at the level of the objectifiable variables. In two of the products (Pampas meat and Amblayo cheese) these changes result from big local changes. In both cases, the products tend to suffer a trivialisation of their quality, that is, the loss of specificity of the quality associated with the area. In the case of the cheese, this deterritorialisation is associated with the partial loss of local know-how. This is seen in the tasting ability of consumers of Salta and at a local level, and in the partial abandonment of natural rennet and the use of modern molds. This tendency is exacerbated by the aging of local population and by a rapid spread of practices that simplify daily tasks.

The cases also show that apart from idealistic issues, subjectivity and cognitive limitations can generate inconsistencies and conflicts in the processes of qualification. The "productive actors" in agri-food systems can produce and elaborate qualities that are substantially different from the expectations of the consumers. Regarding the contradiction between the tendency of consumers to favour natural food and the artificialisation in the production of Pampas beef, Champredonde (2008, 197) postulates the hypothesis that this difference in the definition of quality is due to the fact that the qualities evolved "*...regardless of the wishes of the consumer... given the 'failures' in the translations* (throughout these agri-food systems) *and the interdependencies with other territorial and market phenomena"*. In the case of beef, this situation is aggravated by the fact that the first variable perceived by a consumer of meat is tenderness, and by the difficulty of perceiving changes in the composition of fatty acid in the tasting.

In the case of Colonia Caroya, the main changes were due to the impact of new technologies and the complicity of numerous consumers (mainly extraterritorial) who lack specific tasting skills.

The interaction among objectifiable and subjective variables in the processes of deterritorialisation

On top of consideration of single objectifiable and subjective quality variables, it is also important to consider the interaction between them. We need, on one hand, to observe how the changes at the level of the objectifiable quality influence the transformations of the subjective qualities. For example, the consumers in Salta have got used to eating Amblayo cheese made with chemical rennet, and a part of the new generation of Caroyans have adapted their taste to the fresher "industrial" type of salami. On the other hand, we need to observe how the variables proper to symbolic quality influence the implementation of technical innovations. For example, the practice of peeling and storing cured salami in vacuum packaging is limited, mainly because many consumers consider this a less natural product. The complexity of the interaction between both types of quality variables presents particular points as regards each territorial food. These constitute a spectrum of qualities varying in importance. For Amblayo cheese, a wide spectrum of qualities generated by the diversity in the goat milk production systems, in cheese-making and in market strategy can be observed. The market influences conditions and duration of the cheese ripening, and the variables activated by the process of qualification.

Systems where only goats milk and the traditional mould are used alongside chemical rennet coexist with systems where a mixture of goat and sheep milk but natural rennet are used, etc.. The complexity of the process of qualification implies that in some cases different qualities can be found in the same point of sale, even where all products are called Amblayo cheese.

4.2 Can and should territorial quality be rebuilt?

The evolution of the three foods shows an apparent tendency to de-territorialisation. This general tendency is modified by particular situations specific to each area and production distribution and consumption processes of each food.

Many reconstructions of the situations of the three foods are possible, but the above description raises several urgent questions. These include whether the identity of Argentinean Pampas meat will be lost, whether the quality of Amblayo cheese will be trivialised due to the disappearance of local knowledge, and whether the Caroyan consumer over time will end up adapting to fresh and "industrial" type salami. The three cases show that the processes of territorialisation and de-territorialisation of the quality of food reflect multiple processes in each agri-food system. In fact, they are the result of social consumption. Within overall processes a relative diversity of situations and qualities can be found. The big transformation of Pampas beef quality took place along with economic and social processes of a great magnitude linked to global economic changes such as the expansion of soy crops in direct sowing and the relative pricing of beef, grain and fodder. There was also a background of local phenomena such repeated droughts in some farming zones, financial incentive to slaughter houses by means of differential pricing for feed-lot animals, etc.

There are still production systems with extensive fattening and commercial mechanisms such as the Hilton Quota for beef to export to Europe. In the current context, the reconstruction of territorial quality of these meats would seem possible for small volumes of meat, re-qualified on the basis of symbolic quality variables, such as Geographic Indications, private and collective brands, or the optional labelling in the case of the province of La Pampa (Champredonde 2008). Such tools are more suited to medium to large scale production with the capacity to organise their production systems in response to seasonal demand and quality. The similarity with the case of Amblayo cheese is that big changes are occurring at the level of objectifiable quality variables and hence important territorial transformations. For Amblayo cheese, reconstruction of territorial quality would seem to be possible through collective local initiatives. The challenge would seem to be less difficult than for Pampas beef. For Pampas beef, on the other hand, the challenge to generate a collective brand that contributes to the reconstruction of its typicity would appear to be less difficult. This is because it entails a delimited number of potentially involved producers and there is geographical and cultural proximity among them.

Training in tasting for the people of Salta should also be developed at the same time. Knowing how to appreciate a territorial food allows it to be re-qualified. The adverse factors for this type of projects is, first, the aged local population, and second, the current structure of the market and its mechanisms of qualification. This favour the simplification of production and commercial practices, the image of the product is not affected because demand largely exceeds supply.

Another factor to be taken into account, in a process of increasing formalisation of the commercialisation of cheese, is the adaptation of the premises and tools used in order to comply with health regulations. Today, this adaptation would imply the systematic exclusion of a large part of the local producers. Besides, consideration of this component of objectifiable quality gives rise to the well-known debate on pasteurisation of milk for speciality cheeses. Amblayo is particularly sensitive to this issue as it has a short ripening time. The debate concerns the opposition between legal aspects with the preservation of local qualities based on practices legitimated by the local culture.

In the case of Colonia Caroya salami, the main issue is the controversy between legality and the legitimacy given to certain practices by local traditions. Some producers have adapted to comply with national legislation. They claim to offer quality based on objectifiable variables, such as product safety and the legality of their commercial transactions in procurement and sales. Others seek to confirm their identity as artisan producers, based on practices inherited from their Friulian ancestors. But it is clear that the whole set of local producers looks for the protection of their products under the image a typical product.

Another factor to take into account is that the systematic dissociation of technical, regulatory and objectifiable variables from subjective ones should not lead to regulations that marginalise producers immersed in the local technical culture. The reconstruction of the territorial quality of the Colonia Caroya salami implies, first, the mobilisation of local producers into a project for containment of supply, and codifying various practices and ripening times. Second, although at local level knowledge of tasting is a factor that has kept up territorial quality in local production of salami, it is important to encourage typical salami tasting skills. A preliminary hypothesis can be based on constructivist views, but the basis of this analysis should not be left aside. The quality and qualification of territorial qualities constitute complex social constructions. This involves not only the presentation of possible goals and tools for the re-qualification of the product, but also the consideration of the particularities of the process of construction.

In this sense, new questions are raised regarding for example the governance of systems that will support these collective processes, the competencies of the institutions and professionals that will accompany these processes, and the evolution of the territory and the markets for these products.

5. Conclusions

The contribution of a constructivist view to the comprehension and possible assistance of the processes of re-territorialisation of the food quality allow us to evaluate and mobilise concerns about objectifiable and subjective approaches to quality. Cognitive limitations of the actors directly involved in the agri-food systems are assumed, and may determine important differences and even contradiction in the definitions of qualities throughout systems even at the same stage.

Rationality of the situation allows an understanding of the nature of the processes of quality construction and qualification, and an explanation the unpredictable impact of the transformations of territories. The three cases analysed in this article highlight three aspects to be taken into account: (i) the historic nature of "territorial products", (ii) the process of quality construction by means of interaction between the objectifiable and subjective variables, and (iii) the socio-technological trajectories that result in the current diversity of production systems. These three aspects also explain most of the causes of the territorialisation (de-territorialisation) of each food.

The cognitive limitations of researchers and participants in processes of reconstruction should be also considered. The constructivist view in this sense may frame intervention. Intervention needs to take account of what already exists albeit in an incomplete sense and of the transitory, nature of every process of qualification and observation. The existent and the transitory are both, by definition, dynamic. What exists, exists as a function of interaction between objectifiable and subjective variables in a given context. What is transitory exists as a function of the evolution of work teams, the faculties of its members and of the views mobilised.

So the processes of reconstruction of territorial food quality is an open process of construction, where unforseeable orientation results in complex interactions among the multiple territorial and extraterritorial factors of every agri-food system. The project of re-qualification of food is no more than one of the many factors that interact in this dynamic. The result of the process, in terms of qualification and local impact, is constantly built and rebuilt. It is a process of constant learning in which the goal is to frame this learning within the rules, norms and practices of a given culture that validates them as legitimate. The qualities offered thus constitute an identitary reference for the members of that culture.

Notes

This work received the help of l'Agence Nationale de la Recherche within the program SYSTERRA (ANR-09-STRA-04) and of project DRT-IC of RIMISP.

1. Experiences developed in the Module "*SIAL: investigación/acción en diferenciación de alimentos típicos*" [SIAL: research on the differentiation of typical foods], that belongs to the *Área Estratégica de Teconología de los Alimentos del INTA, Argentina* [Strategic Area of Food Technology of INTA, Argentina]. Investigation carried out with the French-Argentinean Agriterris laboratory and the RIMISP (Latin American Center for Rural Development).
2. In the scientific environment, terms such as *FOSHU, nutraceutic, enriched or functional food*, etc. have been coined (Ruiz Martinez et al. 2003, 467).
3. The author postulates four principles for performance improvement in the line productions: a) the definition of quality agrees with the necessities (of the design), b) the system of quality is prevention, c) a standard management equals zero error, and d) the measure of quality is the price for inconformity.
4. This is one of the multiple definitions proposed by the *Real Academia Española* [Royal Spanish Academy] (www.rae.es/rae.html).
5. The "cost" component in the definition of quality may comprehend also a collective dimension. Japanese engineer Genichi Taguchi introduced the concept of Quality as "*the minor loss possible for the society*". Taguchi used statistics to define the loss function to measure the financial loss resulting from bad quality. (Sukthomya and Tannock 2005).
6. These new products approach to what Salais and Storper (1993) defined as *generic products*, i.e. anonymous as regards to whom they are destined to, and defined independently of the persons.
7. The translation of *boeuf* is beef. However, in France there is the commercialisation under this denomination of meat of diverse category of animals, such as cows, heifers, bulls and eventually calves.
8. Named by Edgar Morin in terms of "neo-archaism".
9. J.-P. Poulain also rescues another phenomenon associated with the symbolic quality of food, the "artistic" innovation of meals, associated with the names of restaurants or famous chefs (1997, 110). This chapter discusses only this one aspect of symbolic quality.

10. Medicine: substance administered with curative or preventive purpose for a disease (Dictionary of the *Real Academia Española*).
11. Cinchón: mold made by a belt of intertwined plant matter. This girth possesses a certain porosity that enables the loss of cheese whey in the period of maturation. A thin porous fabric is placed between the cheese and the container.
12. To qualify: to appreciate, express or determine the qualities or circumstances of a person or thing (www.wordreference.com.ar).
13. This concept is proposed by Laville (2000), based on the contributions of Herber Simon (1959), over the limited rationality of economical actors.
14. Intermediaries that facilitate the sales of livestock by farmers to slaughterhouses.
15. The mixer is a trailer that has internal blades to mix the components of the animal diets (rolls of hay, corn grains, etc.) previous to the distribution to the feeder in the intensive fattening systems.
16. Even one of the enterprises is called *Sotanos de Caroya* (Caroya's basement).
17. Current researches may come up with new elements to deduce the impact of the maturity in the cold store over the flavor of the salami.
18. In relation to the typology made by Oustry as of kitchen elaborations and patio elaborations.

References

Braidot, N.P. (1996). *Marketing Total.* Ediciones MACCHI: Buenos Aires.

Burr, V. (2003). *An introduction to Social Constructionism.* Routledge: London, New York.

Casabianca, F. and Valseschini, E. (1996). *La qualité dans l'agroalimentaire. Emergence d'un champ de recherche*, Rapport final, INRA.

Casabianca, F. et al. (1994). *Maîtrise de la qualité et solidarité des acteurs. La pertinence des innovations dans les filières d'élevage en Corse.* In Cert, M. et al. (eds.), Qualité et Systèmes Agraires: techniques, lieux, acteurs. *Revue Etudes et Recherches sur les Systèmes Agraires et le Développement*, INRA 28: 343-358.

Cazes-Valette, G. (1998). *Anthropologie et comportement du consommateur: le cas de la vache folle.* In Filser, M. and Lemoine, J.-F., *Comportement du Consommateur*, 4-23. Actas de la 2° Jornada de Recherche en Marketing de Bourgogne. Centre de Recherche en Economie et Gestion des Organisations CREGO IAE Dijon.

Champredonde, M. (2008). *Localización, deslocalización, ¿relocalización? de la calidad de las carnes vacunas pampeanas argentina. Impacto territorial.* In Velarde, I., Maggio, A. and Otero G., (eds.), *Sistemas*

Agroalimentarios Localizados en Argentina, 76-99. Buenos Aires: Edición INTA.

—. (2001). *Stratégie des éleveurs et construction de la qualité. Le cas de la filière viande bovine dans le Sud-ouest pampéen, Argentine.* Thesis of doctorate, Toulouse Le Mirail University, Toulouse.

Chávez, M. et al. (2009). *Parámetros de calidad higiénica y sanitaria en leche cruda caprina de sistemas productivos del norte argentino.* In X Congreso Latinoamericano de Microbiología e Higiene de los Alimentos, COLMIC, Uruguay.

Delfosse, C. (2006). *La localisation de la production fromagère: évolutions des approches géographiques. Revista Géocarrefour*, 81 (4): 311-318. http://geocarrefour.revues.org.

Figuie, M. and Bricas, N. (2002). *Risque et qualité des aliments: a la croisée de deux domaines de recherché.* In Actas de 14e Rencontres scientifiques et Technologiques des Industries Alimentaires, Agoral 2002, Prévision, analyse et gestion du risque alimentaire, 251-259, Nancy, 26 et 27 Mars. Paris: Lavoisier, éditions Tec & Doc.

Fischler, C. (1996). Pensée magique et alimentation aujourd'hui. *Cahiers de l'Ocha* 5.

Hirczak, M. (2007). *La co-construction de la qualité agroalimentaire et environnementale dans les strategies de developpement territorial. Une analyse a partir des produits de la Region Rhone-Alpes.* Tesis de Doctorado, Universidad Joseph Fourier.

Juran, J.M. and Gryna, F.M. (1994). *Análisis y planeación de la calidad (3° edición).* Méjico: Ediciones Mc Graw Hill.

Lévy-Strauss, C. (1990) [1962]. *La pensée sauvage*. Paris: Plon.

Mauss, M. (1935). Les techniques du corps. In *Anthropologie et Sociologie.* Paris: PUF.

Muchnik, J., Biénabe, E. and Cerdan, C. (2005). Food identity/food quality: insights from the "coalho" cheese in the Northeast of Brazil. *Revue of* Anthropology of Food, 4 may 2005, Local Foods/ Produits alimentaires locaux, http://aof.revues.org/document110.html

Poulain, J.-P. (1997). *Mutations et modes alimentaires.* In M. Paillat, *Le mangeur et l'animal. Mutations de l'élevage et de la consommation*, 103-120. Paris: Ed. Autrement.

Ruiz Martinez, A. et al. (2003). *Alicamentos y Medicamentos*, 467-470. VI Congreso de la SEFIG y Terceras Jornadas de Tecnología Farmacéutica. Granada, España.

Simon, H.A. (1959). Theories of Decision-Making in Economics and Behavioral Science. *The American Economic Review* 49 (3): 253-283.

Sukthomya, W. and Tannock, J. (2005). Taguchi experimental design for manufacturing process optimisation using historical data and a neural network process model, *International Journal of Quality & Reliability Management* 22 (5) 485-502.

CHAPTER FIVE

IS THE PROTECTED DESIGNATION OF ORIGIN (PDO) POLICY SUCCESSFUL IN SUSTAINING RURAL EMPLOYMENT?

ZORA BOUAMRA-MECHEMACHE AND JAD CHAABAN

Abstract

This chapter seeks to establish whether public agro-food interventions like food quality labels contribute to the promotion of rural employment. To this end, the paper uses original longitudinal firm and plant level datasets on the French cheese industry to assess the impact of the Protected Designation of Origin (PDO) label on rural employment. The data is used to test the impact of the PDO label on equilibrium market structure in the downstream cheese processing industry, and to establish backward linkages this industry has on upstream plant-level employment and the number of dairy farmers. Our results show that the PDO label has increased the equilibrium number of firms at the national level, because it has created market segmentation reducing barriers to entry. In turn, this higher number of cheese firms resulted in more employment in dairy farms and processing plants at the district level. However, the PDO label exerts pressure on farmers to abide by strict production techniques, which may cause exit due to cost increases. Yet our estimates show that the employment benefits of this label outweigh the potential losses it might create due to its product specification stringency.

1. Introduction

The Common Agricultural Policy (CAP) developed by the European Commission emphasizes the role of food quality policy as a tool for product diversification and rural development. It has created different food quality labels, while reducing its direct price support instruments. These quality labels are aimed at protecting the designations of agricultural and food products whose quality comes from their geographical origin. The two main labels are the Protected Designation of Origin (PDO) and the Protected Geographical Indication (PGI) (European regulation 2081/92 followed by regulation 510/2006). Their main objectives are the development of rural activity to increase farmers' income; achieving a balance between supply and demand; developing less favored areas and sustaining rural employment.

The literature on public quality labels has shown that these labels can be efficient tools for producers in signaling the quality of their products. When engaging in a PDO value chain, producers incur additional costs linked to the quality requirements (Bouamra-Mechemache and Chaaban 2010a), but they benefit from a higher quality reputation and thus higher margins. The development of a welfare enhancing PDO label may occur in a perfect competition market as shown by Moschini et al. (2008) in the case where free entry is possible on the market. Yet the profitability of the PDO label for producers ultimately depends on the market structure and the possibility of output control, as shown by Marette and Crespi (2003), Lence et al. (2007), and Hayes et al. (2004).

While the literature on the effectiveness of the PDO label in raising agro-firms' income is abundant (in Bouamra-Mechemache and Chaaban 2010b); to our knowledge the literature which evaluates the impact of food quality labels on the rural economy at large is almost nonexistent. In fact, while the PDO policy applies only to the final food product, the whole production chain (farmers and processors) is involved in its production. This chapter aims at addressing this issue, and contributes to the empirical evaluation of the PDO regulation as a tool for rural development.

It analyses the ability of quality label to sustain the existence of agrofood chain by taking into account its incidence both on downstream firms and upstream farmers. It does not deal with the impact of PDO in the development of tourism in rural areas but rather focuses on employment as a measure of rural activity. More particularly we analyze the contribution of PDO policy to developing employment in the agricultural and food sector especially in less favored areas.

The empirical analysis applies to the impact of the label "Appelation d'Origine Controlée" (AOC) on regional employment in the French cheese sector. The AOC label is older than its European equivalent PDO. France was one of the first countries with Italy to introduce this kind of label. It was first applied to wine in 1935 followed by the cheese sector. All French cheeses produced under AOC label also benefit from the European PDO label. The French cheese sector is thus relevant to analyze the impact of

PDO labels. It enables us to determine the regional effect of the AOC policy on the number of milk producers and on the number of dairy processors.

The empirical assessment of the AOC label requires data on the characteristics of both producers and products. Two databases on the French dairy sector provide this information. The first one (the Annual Firm Survey from SCESS) provides information on firms' characteristics while the second one (the Annual Dairy Survey) provides detailed data on production at the plant level and thus, makes it possible to distinguish PDO from non-PDO production. In addition to this database, we use aggregate data to get information on the evolution of milk producers. We are then able to get information at the district level on a 10 year period (1996-2006).

We test the following empirical assumption by comparing PDO areas (at the district level) to non-PDO ones: the PDO label, by generating product differentiation relaxes the competition between PDO and non-PDO firms. This decrease in competition will lead to a higher number of firms on the market at the equilibrium. This higher number of cheese processors on the market could sustain employment as well as the demand for milk. An increase in the demand for PDO milk in a given area will support the number of milk producers in this area. Our result shows that the PDO label does have a positive effect on the number of milk farmers: a 10% increase in the share of PDO production in a given district would lead to a 8% increase in the number of cheese plants and to a 6% increase in the number of milk producers.

2. Industry structure and market size

The majority of PDO registered products in the EU are processed rather than raw food commodities, like cheese, wine, butter, and processed meat. The development of a PDO label thus requires the participation of both upstream farmers and downstream processors. From a policy perspective, the impact of the PDO regulation on rural development will thus depend on the vertical relations between upstream producers and downstream processing firms. As a result, the outcome of the regulation will not only depend on the upstream market structure but also on the competitive structure of the downstream market. The structure of the downstream market will thus influence both the downstream and the upstream market. This will in turn determine the level of employment both in the downstream industry and in the input procurement market. Next we present one of the main theories that explains the determinants of market structure, and which applies fittingly to food industries.

2.1 Downstream market structure

The determinants of market structure in a given industry can be sketched using the theory developed by Sutton (1991) and (1998) to analyze the effect of market size on the number of firms (the "bounds approach"). Sutton (1991) shows that the number of firms in a free entry market is an increasing function of the size of the market relative to the sunk cost firms have to pay to enter the industry. The larger the market, the higher the number of firms and the higher the sunk costs, the lower the number of firms. The function specifying the lower bound to equilibrium concentration converges monotonically to zero as market size increase. For the type of industries that are characterized by endogenous sunk costs—that is, high level of R&D and advertising investment—Sutton's theory predicts that, even as the market grows without bounds, there is a strictly positive upper bound on the equilibrium number of firms. Empirical studies find evidence of Sutton's predictions (Sutton 1991, Ch. 4, 5), Robinson and Chiang (1996), Lyons et al. (2001), Rogers (2001) and Giorgetti (2003). The market structure will be determined by market size, the level of production economies of scale, entry conditions and price competition. We will test some of the main predictions.

The impact of market size (S) is straightforward from Sutton's theory:

P1. When market size increases, the number of firms in the industry will increase.

The effective market size depends on the industry's minimum efficiency scale (MES). The higher the MES, the lower the number of firms in the industry (Sutton 1991). It turns out that the number of downstream firms increases with the effective market size (S/MES). The MES is also a proxy for technological barriers to entry. If it is high, then large sunk costs have to be incurred by a new firm that wants to enter the market, which ultimately makes its entry less likely. High MES may deter entry and tends to reduce the number of firms in the industry:

P2. The number of firms on the downstream market decreases with the level of the minimum efficient scale (S).

Finally, for exogenous sunk cost industries, Sutton (1991) has shown that any exogenous action that has a positive impact on prices will reduce the toughness of price competition and results in a larger number of firms. So if price competition is softened due to the public PDO regulation, then introducing PDO labels should decrease concentration.

However, the effect of price competition on endogenous sunk cost industries depends on the relationship between endogenous sunk costs and price competition (Symeonidis 1997a; 1997b). If price competition becomes more intense and advertising/R&D expenditure increases (or remains constant), the number of firms will decrease. However, if advertising/R&D decreases, the number of firms may increase or decrease depending on whether the first decrease is sufficient to compensate for the decrease in prices. With PDO regulation, the resulting relaxation of competition may be followed by a reduction in advertising/R&D competition, which makes entry more profitable. But it may also be the case that the PDO regulation results in higher advertising/R&D expenditure by non-PDO firms or by PDO firms that want to differentiate their products from other PDO labeled products. In this case, entry becomes less profitable and the number of firms on the market will not increase:

P3. If advertising/R&D do not increase or do not increase much, we can predict that the PDO regulation will relax competition, thus increasing the number of firms in the industry.

2.2 Vertical linkages

The total number of farmers and plants will increase with the downstream market size, which in turn will depend on the sales to the downstream industry. The number of plants and farmers is directly linked to the number of firms in the market given the vertical structure of a given industry (Fig. 5-1).

Figure 5-1 Vertical structure of the industry and PDO feature

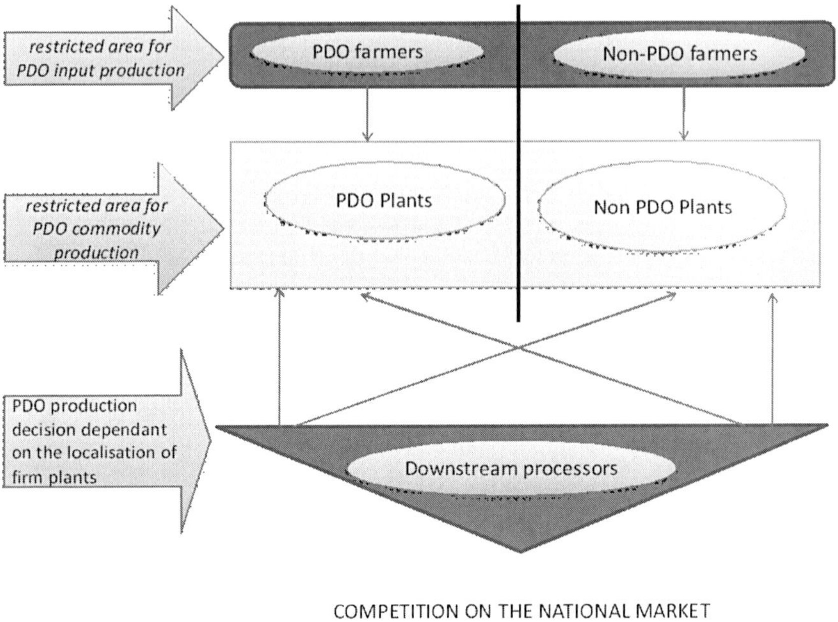

P4. The total numbers of farmers and plants are increasing with the number of firms.

The agricultural market can be seen as a competitive exogenous sunk cost industry with a large number of farmers. First, the number of farmers in a given area will depend on the competitivity of farmers in this area. The relative competitivity of an area will depend on several factors. Among them, climate and geography will similarly influence all farmers in a given area and will also impact the productivity of farmers (production yields and farm size):

P5. The number of farmers is negatively linked with less favoured areas.

In addition, PDO production has a specific feature as both the input and the output productions have to be localized in the PDO region. It turns out that the downstream market features will impact differently the production areas, given the competitiveness of farms/plants in these areas as well as the possibility or not to produce PDO:

P6. The numbers of farmers in a given area are related to the PDO characteristics of this area.

Moreover, the procurement of raw inputs by a plant localized in a given area will depend on the nature of the transaction costs. If the transaction costs increase with distance, then plants have to be implemented in or near the procurement area. For instance, if inputs are perishable or if the transport costs of the input are high, then plants will be located in the neighbourhood of input production units:

P7. The numbers of farmers in a given area is increasing with the number of plants localised in the same area.

In the following section, we test the above predictions for the case of the cheese industry in France. In this industry, the milk produced upstream in the different geographical areas is processed by different plants. Because milk is a perishable product, plants have to be located in the same areas or in neighbouring ones. National firms may have several plants localised in the different areas, but they compete on the national market.

The PDO label may influence milk production activity, as cheese companies have the option to produce either non PDO labelled products, or only PDO ones, or both. Their production decisions and the resulting downstream structure will influence the outcome on the activity in PDO producing areas.

3. Data and empirical model

3.1 Data

We use firm and plant surveys covering the period 1996-2006 provided by the French Administrative Direction of Statistics (INSEE – Institut National de la Statistique et des Etudes Economiques). The first main dataset reports economic and administrative information at the firm level (EAE, Enquête Annuelle d'Entreprise) while the second set reports production activities and provides more detailed information on the industrial process at the plant level for dairy firms (EAL, Enquete Annuelle Laitièere), also provided by INSEE (Institut National de la Statistique et des Etudes Economiques). The EAE dataset is available only for firms with more than 20 employees, while the EAL set is exhaustive at the national French level. The proportion of AOC products in the total production of cheese amounts to around 17% over the period. We focus in this study on AOC cheese made from cow (30 AOC products) and sheep milk (3 AOC), which represent 97% of milk used in the processing of AOC cheese.

Each cheese processing firm is identified by an identifier at the country level (SIREN) and can thus be followed over the entire period. From EAL, we identify all plants using an identifier at plant level (SIRET). This identification number is directly linked to firms' identifier, as the SIRET number is composed of the SIREN number of the firm plus two additional digits that uniquely identify each plant within the firm. We can therefore easily link each firm to its plants. Second, the EAL information at the plant unit makes it possible to have precise information on the production of different types of cheeses. We have defined six categories of cheese products: hard cheese, semi hard cheese, soft cheese, blue cheese, processed cheese and other cheese. For each category, we observe the overall production as well as the proportion of PDO production.[1] Around 90 firms are operational at the national level for each cheese category. From EAE, detailed information on the quantity of milk collected is available. For each plant, we recover the quantity collected as well as the number of farmers by milk production area (at the district level).

From this data, we compute our main dependent variables, which are also interrelated. At the country level, the number of firms per cheese category is denoted by $NFIRMS_{t,s}$ with t is the subscript for year and s the cheese sector category. At the district level (d), the number of plants, plant workers and farmers is respectively denoted by $NPLANTS_{d,t}$, $NEMP_{d,t}$ and $NFARMERS_{d,t}$.

Summary statistics for these variables are shown in Table 5-1. As discussed in the previous section, the main determinants of downstream market structure (i.e., the number of firms) are the market size, the advertising to sale ratio and the minimum efficient sale. These exogenous variables are computed for each observed year at the cheese sector level. The downstream market size (*MSIZEi,t*) is measured as the sum of cheese firms' turnover. The advertising to sale ratio (*ADVi,t*) can be directly computed as overall advertising expenditures per sector divided by the sector's turnover. The French cheese industry could be classified as an endogenous sunk cost industry with a mean advertising-to-sales ratio equal to 6.2%. The Minimum efficient scale per cheese category (*MESi,t*) is computed as the output of the median plant multiplied by the sector's assets. In addition, to evaluate the impact of the PDO regulation, a variable measuring the share of production under PDO label for each cheese category has been computed (*S1_PDO*).

Table 5-1 Summary statistics

Variable	Mean	Std. Dev.	Min	Max
NFIRMS	4.50	0.52	2.71	5.61
NPLANTS	2.72	1.04	0.00	5.59
NEMP	4.81	1.55	0.00	7.88
NFARMERS	6.76	1.16	2.71	9.08
MSIZE	19.48	0.69	17.29	20.28
MES	5.89	0.98	4.61	8.17
ADV	6.15	4.78	0.05	26.01
S1_PDO	0.09	0.06	0.00	0.37
S_MILK	0.40	0.32	0.00	1.00
S2_PDO	0.15	0.25	0.00	1.00
MOUNT	0.19	0.39	0.00	1.00
UNDER	0.28	0.45	0.00	1.00
MOUNTxS2_PDO	0.05	0.15	0.00	0.72
UNDERxS2_PDO	0.03	0.12	0.00	0.72

We use dummy variables to control for the district *k* effect (REG_k). In order to incorporate upstream milk production characteristics, we use a variable measuring the share of PDO production in the district ($S2\_PDO_k$) and dummies that identify districts localised in mountain areas (*MOUNT*) as well as in underprivileged areas (*UNDER*). In our analysis, we also incorporate the effect of milk production externalities on the number of farmers in a given district, by computing the share of milk collected by plants which originates from the same district, as the ratio of plants' overall collected milk input (S_MILK). If in a given district most plants collect milk from farmers located in the same district, then this will have an effect on the optimal number of farmers in this particular district in the long run.

3.2 Empirical model

We use a three-stage least square estimation system to simultaneously evaluate the determinants of the number of firms at the national level, as well as the number of plants, the number of cheese-processing units' employees and the number of dairy farmers at the district level. Equation (1) corresponds to the first market stage where the number of firms in a given cheese sector is estimated as a function of the sector's market size, the minimum efficient scale, the level of advertising and the share of PDO for the sector as well as an error term $\epsilon_{i,t}$.

$$\ln \text{NFIRMS}_{i,t} = \beta_0 + \beta_1 \ln \text{MSIZE}_{i,t} + \beta_2 \ln \text{MES}_{i,t} + \beta_3 \text{ADV}_{i,t} + \beta_4\ \text{S}_1\text{\_PDO}_{i,t} + \varepsilon_{i,t} \tag{1}$$

Equation (2) establishes the direct link between the number of cheese processing plants and the corresponding number of firms at the district level.

$$\ln \text{NPLANTS}_{d,t} = \delta_0 + \delta_1 \ln \text{NFIRMS}_{d,t} + \sum_k \delta_k \text{REG}_k + \mu_{d,t} \tag{2}$$

where $\text{NFIRMS}_{d,t}$ is the equivalent number of firms active in district *d* at time *t*. The equation also controls for regional dummies through the REG dummies.

Equation (3) provides the direct linkage between the number of plants and the number of plant employees at the district level.

$$\ln \text{NEMP}_{d,t} = \alpha_0 + \alpha_1 \ln \text{NPLANTS}_{d,t} + \upsilon_{d,t} \tag{3}$$

Equation (4) establishes the determinants of the number of dairy farmers in a given district. This depends on the number of plants operational in the district and the share of milk from the district collected by these plants, in addition to district-specific dummy variables:

$$\begin{aligned} \ln \text{NFARMERS}_{d,t} = {} & \theta_0 + \theta_1 \ln \text{NPLANTS}_{d,t} + \theta_2 \ln \text{S\_MILK}_{d,t} + \theta_3\ \text{S}_2\text{\_PDO}_{d,t} \\ & + \theta_4 \text{MOUNT} + \theta_5 \text{UNDER} + \theta_6\ \text{MOUNT*S}_2\text{\_PDO}_{d,t} \\ & + \theta_7\ \text{UNDER*S}_2\text{\_PDO}_{d,t} + \omega_{d,t} \end{aligned} \tag{4}$$

We estimate two systems of equations by applying Zellner's Seemingly Unrelated Regression (SUR) technique. We first estimate (1), (2) and (3) jointly to analyse the determinants of industry employment per district. We then focus on the number of farmers by estimating the system of equations (1), (2) and (4). Econometric estimations are conducted using Stata 10.

4. Results

Table 5-2 details our first set of regression estimates. Our sector-level regression confirms our first and second predictions above (P1 and P2).

Table 5-2 Three-stage Least Squares estimation results, determinants of district-level employment in dairy processing plants

	NFIRMS coef	NPLANTS coef	NEMP coef
MSIZE	0.384***		
MES	-0.234***		
ADV	-0.024***		
S1_PDO	3.237***		
NFIRMS		0.405***	
reg (dummies)		(omitted)	
NPLANTS			0.892***
_cons	-1.678**		2.407***
R-sq	0.44	0.91	0.29

Nb. Obs. 337

Note: *** p<0.01, ** p<0.05, * p<0.1

The number of cheese processing firms on the downstream market increases with market size (*S*) and decreases with the level of the minimum efficiency efficient scale (*S*), as both parameters for MSIZE and MES are highly significant with the expected signs. Interestingly, our third prediction (P3) is also validated by our estimation results, as the number of firms seems to decrease with sector's advertising intensity, and to increase with the share of PDO production in the sector. In any given cheese sector, a high PDO production seems to promote the number of firms operational on the market.

As for the other regressions, the number of firms coefficient has a positive and significant effect on the number of plants, and the number of plants in turn have a positive and significant effect on the number of plant employees (these are expected direct linkage effects).

Figure 5-1 Number of firms in relation to Minimum Efficient Scale, by PDO intensity

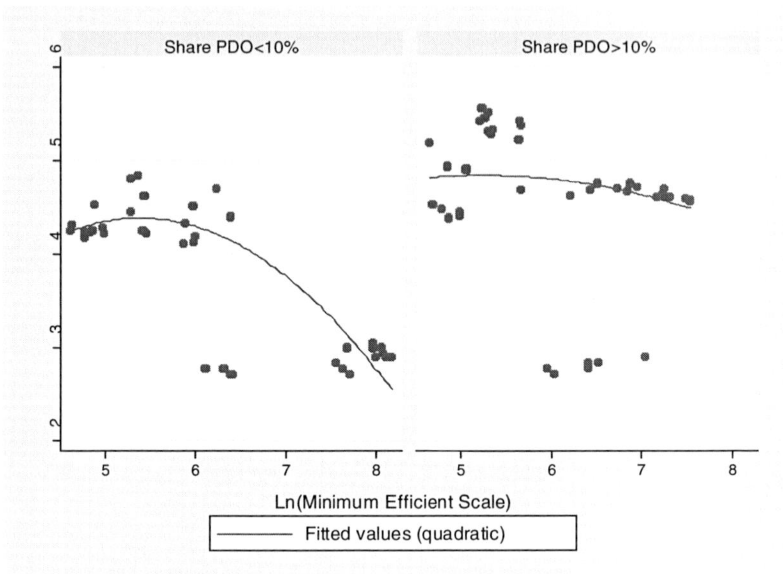

The effect of Minimum Efficient Scale (MES) on market structure (the number of firms) is detailed in Fig. 5-1. When the MES is low (indicating low barriers to entry), the number of firms increases slightly. When the MES becomes higher, technological entry barriers increase, resulting in a lower equilibrium number of firms. However, the impact of the MES follows a different pattern for non-PDO and PDO intensive sectors. While the MES has a strong negative impact for non PDO firms, it marginally affects PDO firms, suggesting that PDO firms are not hurt as much as non-PDO firms by market barriers. PDO regulation relaxes competition by compressing the production scale necessary to be profitable under this label.

Table 5-3 provides estimates of the second set of equations, where we now include now the number of farmers at the district level. Similar to the previous results, the regressors in the NFIRMS and NPLANTS equations are all significant with the expected signs. Also, as expected, the greater the number of plants in a district and the higher the share of milk collected from within the district increases the number of dairy farmers. Interestingly, the higher the share of PDO production in the district

(S2_PDO) reduces the number of farmers there, but not when crossed with districts that are in mountainous areas.

Table 5-3 Three-stage Least Squares estimation results, determinants of district-level numbers of dairy farmers

	NFIRMS coef	NPLANTS coef	NFARMERS coef
MSIZE	0.395***		
MES	-0.238***		
ADV	-0.025***		
S1_PDO	3.086***		
NFIRMS		0.531***	
reg (dummies)		(omitted)	
NPLANTS			1.418***
S_MILK			0.609**
S2_PDO			-0.905**
MOUNT			-1.810***
UNDER			-1.613***
MOUNTxS2_PDO			1.652**
UNDERxS2_PDO			-0.144
_cons	-1.855***		12.656***
R-sq	0.45	0.91	0.53

Note: *** p<0.01, ** p<0.05, * p<0.1.
Nb. Obs. 349

5. Conclusion

The empirical analysis has showed that the PDO label seems to positively contribute to the sustainability of rural employment, mainly through the downstream market channel. A higher share of PDO cheese products relaxes market competition, which enables more firms to operate in a larger market. In turn, a greater number of cheese processors results in more processing plants, typically located close to dairy farmers, which then translates into greater (or sustained) employment in these plants and higher numbers of dairy farmers.

However, the PDO label seems to have a direct negative economic influence on the number of farmers at the district level. Our econometric results show that the number of farmers decreases the higher the share of PDO at the district level. This seems to indicate that the PDO label is imposing constraints on farmers by stringent production technique requirements, which ultimately impacts their production costs. This negative effect is however compensated for by the downstream market effect, where demand for milk from plants affiliated with cheese processors is expanding because of the impact the PDO label exerts on the overall market structure downstream.

Notes

1. PDO can be directly observed for most of PDO cheese. For other PDO cheese, we recovered information at the plant level using other source of information. It was not possible to recover the information for goat cheese.

References

Bouamra-Mechemache, Z. and Chaaban, J. (2010a). Determinants of Adoption of Protected Designation of Origin Label: Evidence from the French Brie Cheese Industry, *Journal of Agricultural Economics* 61 (2): 225-239.

Bouamra-Mechemache, Z. and Chaaban, J. (2010b). Protected Designation of Origin Revisited. *Journal of Agricultural & Food Industrial Organization* 8 (1), Article 5. Available at: http://www.bepress.com/jafio/vol8/iss1/art5.

Giorgetti, M.L. (2003). Lower Bound Estimation: Quantile Regression and Simplex Method: An Application to Italian Manufacturing Sectors. Journal of Industrial Economics 51 (1): 113-120.

Hayes, D.J. et al. (2004). Farmer-owned brands. *Agribusiness* 20 (3): 269-285.

Lence, S. et al. (2007). Collective marketing arrangements for geographically differentiated agricultural products: Welfare impacts and policy implications. *American Journal of Agricultural Economics* 89: 947-963.

Lyons, B.R. et al. (2001). Industrial Concentration and Market Integration in the European Union. *Economica* 68 (269): 1-26.

Marette, S. and Crespi, J.M. (2003). Can quality certification lead to stable cartels? *Review of Industrial Organization* 23 (1): 43-64.

Moschini, G. et al. (2008). Geographical indications and the competitive provision of quality in agricultural markets. *American Journal of Agricultural Economics* 90(3): 794-812.

Robinson, W. and Chiang, J. (1996). Are Sutton's Predictions Robust? Empirical Insights into Advertising, R&D and Concentration. *Journal of Industrial Economics* 44(4): 389-408.

Rogers, R.T. (2001). Structural Change in US Food Manufacturing, 1958 to 1997. *Agribusiness* 17 (1): 3-32

Sutton, J. (1991). *Sunk costs and market structure: Price competition, advertising, and the evolution of concentration*. Cambridge, Mass. and London: MIT Press.

Symeonidis, G. (2002), *The Effects of Competition: Cartel Policy and the Evolution of Strategy and Structure in British Industry*, vol. 1. Cambridge, Mass. and London: MIT Press.

CHAPTER SIX

THE CRAFT VILLAGES OF THE RED RIVER DELTA (VIETNAM): PERIODIZATION, SPATIALIZATION, SPECIALIZATIONS

SYLVIE FANCHETTE

Abstract

In the densely populated countryside of the Red River Delta (1,000 inhabitants per square kilometre), an original, local production system is in the process of being modernized and expanded. Founded on the bedrock of craft and industrial villages connected within commercial and family networks, clusters of craft villages seek their way in the context of the ongoing transition to a market economy of a system long administered by the communist State.

A pre-capitalist system of industrial production, craft industries, organized into clusters in Vietnam, have not yet been swept away by capitalism, unlike in other East and Southeast Asian countries, dominated by the market economy and large-scale industrial complexes that employ a numerous, low-paid workforce. In the shadow of China, whose companies are difficult to compete with, Vietnam manages to carve out a niche and continues to produce artisans. But for how long?

Drawing on a diachronic study, we will present the methods of economic restructuring and spatial reform of various cluster types (textiles, woodwork, food processing, wickerwork and metallurgy), the technical transformations of various crafts and their workforce capacity, according to policies implemented and local, national and international economic circumstances.

1. Introduction

In the densely populated countryside of the Red River Delta (1,000 inhabitants per square kilometre), an original local production system is in the process of being modernized and expanded. Founded on the bedrock of craft and industrial villages connected within commercial and familial networks, clusters of craft villages seek their way in the context of the ongoing transition to a market economy of a system long administered by the communist State.

Craft villages emerged as early as the 11th century in the environs of Hà Nôi for the production of objects necessary to the workings of the imperial capital and to supply products for the everyday needs of a village society geared towards home consumption. Since time immemorial, they have been tied into networks of age-old connections, linked to the capital by the "District of the 36 streets and guilds" and to areas where raw materials are extracted, to markets both national and international (principally China). Since the Economic Liberalization of the 1980s, or *Doi Moi*, growing numbers of villages have developed fresh activities, or have intensified their production. This thanks to the opening up of borders, the freedom to do business, the acceleration of town-country links which were restricted during the collectivist era, and thanks to national and provincial incentive policies.

Currently, there are about 1,000 craft villages in the Red River Delta organized into clusters. These localities produce articles designed for the daily life of villagers (food processing, religious objects, industrial products and building materials, commercial services and transport…) and for export (wickerwork, furniture, woollen clothing and *objets d'art*…). They enjoy income several times greater than that of "agricultural" villages and have thus been able to invest in real estate and improve their living conditions.

In terms of production growth, employment capacity of a plentiful underemployed workforce in the countryside and of the resilience of this type of industry during the industrial crises of the collectivist period, craft villages have proved their worth.

As we contemplate the history of handicrafts in the Red River Delta over the course of several centuries, an endless cycle seems to emerge: its narratives tell of the birth of the crafts, their dissemination and specialization, technical improvements and eventually their death and subsequent resurrection.

The type of cluster organization characteristic to Vietnamese handicrafts is a pre-capitalist system that has survived the advent of the market economy. This stands in sharp contrast with the rest of East and Southeast Asia, where large, international corporations thrive on laissez-faire economics and cheap labour. Despite being in the long shadow of China, whose businesses are hard to compete with, Vietnam is managing to steer its own course and continues to produce artisans. But for how much longer?

2. Craft villages: an ancient, territorialized production system in perpetual reformation

Linked closely since their creation with the capital Hà Nôi and a dense network of markets (Nguyên Duc Nghinh 1993), craft villages have been organized into clusters, both for production and for marketing. These villages were linked by a latticework of canals and rivers to Hà Nôi, the city located "within the bend in the river", constituting a veritable communications hub between the mountains and the China Sea, opening out to the world.

This production system is specific to densely populated rice-growing deltas. In these Asian monsoon plains, replanting work, a technique necessary for attaining very high rice yields and obtaining two harvests every year, has seasonally required a large workforce. Rice, a colonizing plant, has the merit of feeding a large population, but demands intensive bursts of work at specific times of the year.

During the off-season, labourers have had to find other activities, their small plots of land being insufficient to feed them all year round. In this manner, several villages on these overpopulated plains have specialized in other non-agricultural activities that require little capital outlay and are capable of temporarily absorbing a large workforce.

2.1 Craft villages organized into clusters

Craft villages began to thrive as the Chinese conquest was ending. After Emperor Ly Thai Tô moved the Imperial Court from Hoa Lu to Thang Long on the banks of the Red River in 1010, several villages began developing craft activities. They were granted monopolies and dedicated themselves fully to the manufacture of religious and luxury handicrafts for the Imperial Court and the affluent classes in Vietnam and other Asian countries (such as China and Japan).

Craft products were also used in the daily life of the villages, which were largely geared towards home consumption: cotton cloth, ceramics, agricultural and hydraulic tools, rattan and bamboo goods, sedge mats, processed foodstuffs (e.g., from flour mills, oil mills and distilleries) and industrial products (metalwork, milled paper) (Fanchette and Stedman 2009). A wide range of items could be manufactured, owing to the availability of a large variety of botanical and animal raw materials.

Each village produced infinite variations on the one item it specialised in. Wickerwork is the best example of this diversity: there are eight varieties of bamboo, each with its specific characteristics, from which an endless variety of baskets can be made: baskets of all shapes and sizes, with a more or less tight weave, baskets for cooking, for transporting the harvest or soil, for irrigation, for drying the harvested crops, for preserving goods or breeding silkworms; watertight baskets to carry water. In 1930, Pierre Gourou counted more than 800 crafts in the Red River Delta (Gourou 1936).

This same author observed a tendency for each village to specialise in one specific activity and for labour to be divided among villages, which thus become integrated production groups manufacturing the same type of goods. This development probably started at the onset of the 17th century, at a time when handicrafts underwent rapid evolution stimulated by the growth of Hà Nôi. Aspects of the burgeoning production process called for rationalization, including:

- A monopolistic mindset (organized around social rules and rituals): once they had succeeded in developing an industry, villages wanted to protect their production methods, which therefore had to be kept secret.
- Poverty meant that craftspeople sought to make a swift profit and lacked the means to purchase large amounts of raw materials or otherwise tie up capital. Individual villages would therefore only carry out part of the production process and sell semi-manufactured goods on to another village, which would in turn manufacture the finished product.

This localized system of production is founded on villages, the territorial and social building blocks of Vietnamese society. They are communities of shared skills within a territorialized and demarcated space defined by an architectural and religious heritage consisting of communal houses, temples or small pagodas designed for the worship of the village's craft ancestors and tutelary spirits. A rich, intangible heritage finds expression in these places and bears witness to the defining role of these crafts within this village society (Fanchette and Stedman 2009).

2.2 Villages linked to the capital through the "District of 36 streets and guilds"

Traditionally, cities in Vietnam were symbols and seats of power and were regarded as religious centres. They were military strongholds responsible for defending the integrity of the national territory and the place of residence of the celestial appointee—king or lord—or his representatives (Langlet Quach Thanh-Tâm 1993). At the same time, they were also trade centres.

During the Ly Dynasty (11-14th centuries) communities of artisans working in the same craft gradually moved to Ke Cho, the capital's market. They set up shop in "urban villages", while continuing to go back and forth to their native villages. This constant to and fro movement stimulated trade and the craft industry. The city workshops called upon the vast reserves of rural manpower that contributed to the enrichment of this district and of their villages: orders placed in the capital provided work for the craft villages, the city representing an immense market for these products, some of which were made there on the spot.

Then with densification and urbanization, what had once been large urban villages specialized in a one craft activity turned into specialized streets. There were in fact roughly one hundred streets – 36 was chosen as an auspicious number. Each street was named after the goods it sold: Sugar Street, Hemp Street, Cards Street, Dyers' Street, Cup Street, Cotton Street, Chicken Street, Tray Street, Tin Street, Drum Street, Fan Street, Comb Street… (Papin 2001). Shopkeepers were sometimes themselves artisans. Trade with foreign countries stimulated craft industry and the city was visited by Western and Chinese merchants (Nguyên Thua Hy 2002).

2.3 A city at the crossroads between markets, which lost its role during the collectivist era

Thanks to its favourable location at the crossroads of multiple waterways, the Red River, the Tô Lich River and the Kim Nguu River (Hà Tây), the "District of 36 streets and guilds" was at the heart of regional commerce. Via numerous ports located along the Red River and the Tô Lich River, merchandise from craft villages in the Delta was brought to the specialized streets of the capital.

City markets were held at the various gates of Ke Cho's surrounding walls. There were eight large ones in the 18th century, excluding specialised markets such as the rice market, fish market and frog market. Small markets would take place in front of the Red River wharf; brine, salt and sugar were traded in the streets bearing their names, which are located near the river (Nguyên Thua Hy 2002). Markets specialized in craft products, silk or paper, for example, were held in the "District of 36 streets and guilds" or round West Lake, on the outskirts of the capital.

Then, in the 1950s, during the collectivist era, private enterprise was outlawed. Artisans from the villages were organized into production co-operatives. In the more active villages and where local authorities were more "understanding", some artisans were able to continue production clandestinely. Purchase of raw materials became difficult and selling craft produce called for a lot of imagination (DiGregorio 2001). The handicrafts industry then experienced two contradictory trends: production was reduced in certain sectors as a result of the ban on individual production, while it increased via craft co-operatives in sectors under tight State control. Craft co-operatives were to fulfil several functions:

- Enable provinces to become self-sufficient in agricultural and hydraulic machinery and tools in order to increase agricultural production and manufacture supplies for everyday life, such as paper, furniture, pottery, food products, etc.
- Produce commodities for Eastern European countries—the West was in the grip of the Cold War and therefore boycotted the new Communist State of Vietnam.

Trade was controlled by State enterprises in the "District of the 36 streets". It lost its role as intermediary with craft villages. After the economic liberalization of the 1980s, only a few streets managed to re-establish these ties, but as markets have developed for export or towards the provinces in the south of the country, fresh trade networks have been established.

2.4 A growing number of craft villages since Doi Moi

After the fall of the Berlin Wall, co-operatives could no longer rely on the preferential markets of Eastern Europe. After decades of State control, craftspeople were faced with a tough challenge: in the context of Dôi Moi and its open-door policy, they would either have to find new markets by themselves or stop production. Some villages managed to adapt to the new context, depending on the nature of their activity and on craftspeople's personal history and social networks (especially relations with the merchants of Hà Nôi). Today, some craftspeople run prosperous international businesses, some have reinvested their talents in other commercial or production activities, while others have left the village or gone back to agriculture.

Some of the more active craft villages developed private initiatives long before the advent of Dôi Moi, thanks to the acquiescence of local authorities aware of the failure of the collectivist system. A number of disgruntled craftspeople left dysfunctional co-operatives and continued practising their craft illegally (DiGregorio 2001). When economic liberalization occurred, they were ready to integrate the market economy system.

Although a certain number of villages lost their craft (weavers, itinerant carpenters), or switched to fresh activities (trade, food processing), others mechanized and increased production. Some clusters expanded with the opening up of foreign markets, particularly those producing wickerwork and fine furniture. The most dynamic villages shared their techniques with surrounding communities. More than a quarter of craft villages in Hà Tây Province developed craft activities with the advent of Dôi Moi.

Thus in the year 2006, there were about 1,000 craft villages in the Red River Delta, employing roughly a million workers.

In Hà Tây Province, a few large clusters of more than 10,000 workers emerged: La Phu (knitting and confectionery), Huu Bang-Chang Son (mid-range furniture), Duong Liêu-Cat Que - Minh Khai, an important centre for the processing of agricultural produce (noodles, glass noodles, starch…). As for the host of villages producing wickerwork, they became organized more in networks than in clusters, but employ many workers through subcontracting on a temporary basis.

In Bac Ninh, the most dynamic cluster of villages is Dông Ky (fine furniture) composed of three contiguous communes that employ 25,000 people, of which more than a third are workers from outside the cluster. Da Hoi is at the centre of a cluster of nine villages specialized in the production of metal bars and rods for the construction industry. In 2006, the commune was home to about 5,000 artisans and workers, and employed around 7,000 others from outside. Finally, a smaller, heavily industrialized cluster specialized in papermaking covers the commune of Phong Khê with the village of Duong O at its centre, and an industrial site established within the commune.

3. A local production system that integrates all the stages of production and marketing

In spite of the upheavals of the economic and political history of the last century, craft villages have managed to survive and even to prosper. Our hypothesis is that the resilience of craft villages can be explained by their bunching into clusters, a very flexible organization of villages in terms of use of workforce and capital and within which the production process is fragmented and the economic risks spread widely.

3.1 The organization of clusters

The organization of clusters takes place on three levels (Fanchette and Nguyên Xuân 2009):

- Between villages:

A cluster of villages is composed of a central village (the mother-village) and of secondary villages. At the centre are the major producers and placers of orders, official businesses, often mechanized and concentrated within a designated craft park. Several types of inter-village relations can be found within clusters. Each village specializes in one type of product but relies on others for:

- Supply of raw materials (trade, sorting, recycling): for villages using recycled products, there is a long processing chain for these materials. Workshops that carry out this kind of work generally belong to secondary and poorer villages of the cluster that have a numerous workforce and scant means to set up mechanized workshops.
- Knowhow: some villages have had a craft specialization for centuries and take part in the making of luxury articles or objets d'art (woodcarving, mother-of-pearl inlaying, joinery…). These are often villages where the artisans are not businessmen or salesmen and rely on more dynamic villages to market their products.
- Production space: large-scale contractors in the mother-villages have big workshops and, lacking sufficient land in their villages, they rent plots of land in neighbouring villages.
- Services: (transport, trade, markets for raw material, workshops for repairing machines) and other related activities linked to the principal activity (packaging, dyeing or sewing services for textile villages).
- Supply of manpower: the extension of the employment area of clusters takes place in two directions: the closest communes for subcontracting and the hiring of some temporary workers, and the provinces on the fringes of the Red River Delta for longer-term workers, without experience, but working for lower wages.

- At village level:

Work is divided up between complementary businesses, each responsible for one stage of production, or which may also specialize in one type of product. Mechanization and diversification have extended the production chain and generated an increase in the division of labour between households. Recycled raw materials (paper or metal) are traded down a long line of collectors, then processed by craftspeople who have invested in machines-smelters sell recycled metal in the form of ready-to-use sheets, which can then be cut up to make pots, trays and gongs.

- Between village businesses and formal sector companies on industrial sites:

Large-scale companies established on urban industrial sites subcontract the production of spare parts to specialized workshops in craft villages. This type of relationship can be found in the metalworking industry.

3.2 The formation of clusters

Villages organized themselves into clusters several centuries ago to respond to the fragmented organization of work (and thus avoid competition), and for specialization and optimum use of raw materials.

Division of labour is particularly noticeable in the silk industry, where the production process entails several different stages. Each stage is carried out by a different village within a network. Some activities were complementary: the lacquerware manufacturers of Bing Vong often worked with woodcraft or wickerwork villages.

Another factor explaining the high degree of interdependence among villages is the fragmented way in which raw materials are used for the manufacture of various goods (e.g., bamboo or latania palm leaves). Each village only uses part of the raw materials for the manufacture of the goods it specializes in, and then sells on what is left to other villages, which use it to produce the category of objects they specialize in.

During the collectivist era, several crafts were spread through co-operatives. As the taboos concerning trade secrets no longer held, co-operatives specialized in certain kinds of artistic products formed large groups of artisans capable of delivering orders to Eastern European markets. In the villages around Phu Vinh, a famous village specialized in rattan weaving, training courses were organized. The craft was shared with about ten surrounding villages. Today, rattan and bamboo weaving has become the speciality of craft villages in the district of Chong My and employs nearly 35,000 villagers (Fanchette and Stedman 2009).

The spread of new crafts has taken place around the most dynamic villages since Doi Moi. Villages having undertaken mechanization and modernization of their activity have scaled up and diversified their production (Fanchette 2007); businesses need vast workshops and employ a large workforce (either *in situ*, or subcontracted). In order to respond to a growing market in the construction industry, the metalworking village of

Da Hoi set up a 13 ha industrial zone at the beginning of the 2000s, where blast furnaces have been installed. Businesses that were unable to obtain land here built workshops in surrounding agricultural villages or on the small industrial park in the commune of Dinh Bang located along the main road. Fresh crafts have appeared in the poorest villages of the commune: couriers in the village of Da Van link up the many businesses in the cluster, particularly those enclosed within the congested village centre. In this same village, people weave wire to make anti-erosion devices for dykes or chicken baskets and cages, and all other kinds of utensils for daily life. Shops selling metallic bars and rods open in neighbouring villages, while others provide workers for businesses in Da Hoi.

3.3 Conflicting new trends in the development of clusters

Two development processes of clusters can be witnessed with the mechanization and formalization of a growing number of businesses working for export.

The extension of the production chain, with fragmentation of the process is as follows:

- Each machine must pay its way; artisans specialize in a single stage of mechanized production.
- Specialization of artisans in sophisticated techniques.

A larger number of workers are involved in the production process and seek to diversify their activities. The cluster takes up more space and opens up to other markets. The example of the aluminium beaters who have mechanized the pressing of receptacles is significant: they produce them more cheaply, in greater quantities, and in a more standardized manner than those who work by hand.

The production chain for fine furniture is made up of eight stages: from cutting up the wood, first treatment of these pieces, carving, mother-of-pearl inlaying or varnishing, through to assembling. These specialized workshops have a geographic dimension that depends on both their needs in terms of production space and the harmful effects that they generate (Hamel 2010).

However, lengthening the production chain makes it harder to control the quality of items produced. It is not possible to impose quality labels and establish strict specifications for the production process. Delivery deadlines for orders are difficult to comply with because between each stage, whether it be for items made by subcontracting or all along a chain of producers who sell partially made items on to the next artisans, time is continually being lost. These two factors do harm to the export market. Wickerwork, mostly destined for export, is undergoing an acute crisis because foreign markets are much harder to satisfy. The lengthening of the chain weakens relations between the members of the cluster inasmuch as a greater number of artisans take part in it. Familial and neighbourly relations no longer suffice to guarantee confidence in exchanges and services. Fresh ties are developing between those who place orders and the subcontractors: owning a machine that carries out a specific stage within the chain is becoming a reason to develop links between artisans (Hamel 2010).

- Control of the whole of the production chain by big businesses: clusters "pulled along" the commodity chain by traders.

Mechanization, which goes hand in hand with the modernization of formal businesses that seek to standardize their production and speed it up (Fanchette 2007), is moving in the direction of greater control of the entirety of the production process and reducing the subcontracting of production stages.

This takes place in two kinds of sector:

- Very mechanized ones: industrial papermaking. The materials used are recycled paper or imported cellulose pulp. Industry has replaced craft, but producers, in order to avoid competing with industries in the modern sector (joint ventures or State-owned enterprises that enjoy numerous fiscal advantages) draw upon the flexibility of the cluster. They have a foot in each system and, depending on the season, seek to supply various kinds of markets.

- Wickerwork and lacquer businesses employ numerous artisans and subcontracted workers. The production chain is fragmented and the various stages, some of which can only be done manually (weaving), are carried out in villages that are sometimes very remote. The bosses of large businesses themselves try to control the whole production process. They create workshops in various places where they manage to rent land and subcontract to an expert workforce. They can control the purchase of types of bamboo and their treatment on the spot, thus avoiding dependence on intermediaries. These kinds of companies, aimed at export markets, are in effect run by traders who place orders with heads of production groups. Not emanating from the cluster, these businesses set up shop near reserves of manpower to take advantage of knowhow, modest wages and low rental prices.

The rate at which villagers are renewing and replacing craft activities is indicative of these people's potential for adapting rapidly to new situations. However, it also highlights the frailty of certain crafts that are particularly vulnerable to economic and social conditions or suffer from geographical isolation.

The factors for change within this country in transition towards a market economy are the result of both the effects of many decades of a collectivist system (management methods, workforce organization, kinds of market, absence of initiative) and those linked to liberalization (the opening up of the property market, individualism, diversification of production, mechanization, disengagement of the State and competition from foreign countries, mainly China). After a decade where craft villages have had to fight for survival, adapt to new markets and innovate, a natural selection is taking place among the various kinds of activities: some are expanding, while others are disappearing. Many craft products designed for everyday life have been replaced by Chinese items produced industrially and sold at very low prices. Itinerant carpenters, *the môc,* have vanished from the landscape: large-scale water management projects carried out during the collectivist era have made biannual or even triannual rice harvests a reality, and the concrete strip houses that replace traditional dwellings with heavy wooden beams have turned this craft—formerly organized into guilds—into a thing of the past. Some *the môc* villages turned to the production of rattan and bamboo goods, which requires little investment, or gave up crafts altogether.

In general, therefore, a halt in production of a particular kind of item does not necessarily mean the demise of craft activities in a village: artisans retrain for more profitable activities and learn fresh crafts.

Organized into businesses or companies, more occasionally into co-operatives, dynamic artisans amaze observers through their capacity to connect with markets, with new techniques and to follow the State's modernization policy. The cluster system survives and adapts more easily to the current economic climate than State or private enterprises in the modern sector. However, the negative aspects of this rural industrialization are numerous, and particularly concern environmental and social matters (working conditions of employees unprotected by Employment Law are very tough).

4. Spatial history of two clusters with contrasting destinies: silk weaving and fine cabinetmaking

The seniority of craft village clusters is not a guarantee of longevity. Thousand-year-old villages that still have a reputation are above all specialized in *objets d'art*, woodcarving, metalbeating, etc. Textiles, in particular silk weaving—although famed well beyond the national borders at the time of feudal state and employing a large number of craftspeople less than a century ago—has seen production fall steadily since the collectivist era. On the other hand, a recent craft such as that of fine cabinetmaking owes its success to a small group of very dynamic former buffalo traders that managed to establish a network of traders and artisans whose influence spread well beyond borders.

4.1 Silk weaving: an ancient craft with a very fragmented production process

During the colonial era, the textile industry was preponderant in Hà Tây Province to the west of the capital: it accounted for 20,000 artisans out of a total of 54,000 in the Delta. They were to be found in 37 villages specialized in cotton fabrics and their by-products and 25 of silk spinners or producers of silk fabrics (Gourou 1936). These villages were grouped into two clusters not far from Hà Nôi, seat of political power until the 19th century and the biggest market for the consumption of luxury goods:

- Near Hà Dông was the silk cluster composed of seven villages all with La in their name (La Phu, La Khê, La Nôi, La Duong, La Ca, Dông La, Y La) and 3 with Mô. The other villages were located to the north and south of this major textile centre.
- The cluster around the shores of the great West Lake: where five villages were specialized in silk gauze in the 17-18th centuries.

Division of labour is particularly noticeable in the silk industry, where the production process entails several different stages. Each stage is carried out by a different village within a network: growing mulberry trees and breeding silkworms, reeling off the silkworm's cocoon, spinning silk, weaving and dyeing. Many weaving villages were unable to breed silkworms, because their land was not suitable for growing mulberry trees. Only villages located on the alluvial banks of the Day River (unprotected by the dykes) or on the shores of the great West Lake engaged in this activity. In addition, each weaving village specialized in one of the many different varieties of silk – organza, brocade, taffeta, velvet or raw silk.

During the colonial era, two kinds of industry coexisted: a purely familial industry, traditional, making rough material, belts, the strings for simple tops for women, and a modernized industry, often employing a salaried workforce and producing finer materials, sometimes indistinguishable from Chinese or European ones. Birthplace of silk weaving in Hà Dông, Van Phuc is atypical among most of the Delta's craft villages, with capitalist-style production systems, regular employees and the mechanization of production (200 Jacquard looms imported from Lyon). These workshops were more prosperous than purely familial and traditional enterprises. Van Phuc, in keeping with La Khê, its neighbour, was home to a large number of famous artisans. Its export markets spread to Asia, then during the colonial era to France.

The weavers were specialized in a single kind of silk: that designed for the working classes and coarsely woven, more precious cloths such as *lint* satins or brocaded *gâm* designed for the Imperial Court and for rich traders. There was a large variety in qualities of silk, each being the speciality of one village (Nguyên Thua Hy 2002).

These villages have developed according to fluctuations in the market and in fashions. Western travellers and missionaries who crossed this region up until the 19th century never tired of praising the quality of Hà Noi silk. When French colonization occurred in the 19th century, coarser materials suffered from the competition of cotton fabrics imported from France.

Today silk making has almost disappeared from Hà Tây. The collectivist interlude killed off this activity following the suppression of luxury markets, a ban on silkworm production and the reorganization of artisans into co-operatives. Weavers therefore had to switch to the production of low-quality cotton fabrics on wider looms in co-operatives controlled by the State to supply the country at war. Attempts to restart the activity after Dôi Moi have been unsuccessful, owing to the competition from Chinese products, the use of silk threads mixed with synthetic fibres and the absence of quality control.

Of the twelve villages in the Hà Dông silk cluster, Van Phuc is the only one to have maintained the craft despite the upheavals of history. A few talented artisans from La Khê have unsuccessfully tried to rekindle the craft within a co-operative. La Noi, La Duong and Y La produce mediocre cotton fabrics for the domestic market, while La Phu has specialized in knitted garments.

These cotton fabric and knitted garment-producing villages are also organized into clusters, but they answer to industrial commodity chains in China or in the south of Vietnam.

4.2 Dông Ky: a recent and very dynamic fine cabinetmaking cluster

This cluster of craft villages emerged at the end of the collectivist era to become one of the most dynamic in the environs of Hà Nôi. In contrast to the ancient silk cluster, it owes its success to the spirit of enterprise of a handful of traders, who belatedly became artisans, and built up a veritable international network of producers of fine furniture and traders in wood.

Dông Ky enjoys many economic and demographic advantages:

- A community of buffalo traders who switched to other activities, once agriculture became mechanized in the co-operatives, and used their capital and their trade networks to find supplies of wood in the mountainous regions and market their products as far away as the south of Vietnam.
- A spatial and demographic dominance compared with the other villages of the cluster. There were about 12,500 inhabitants in 2006, compared with 2,000 or 3,000 inhabitants in the villages of neighbouring communes.
- A favourable location near major travel and trade corridors.
- A neighbourhood of villages specialized for several centuries in woodcarving and the building of pagodas, temples or communal houses.

Originally, this village had several crafts, including coarse weaving practised by women. Artisans built the frameworks of houses and public buildings, made simple furniture and carried out orders placed by State trading agencies.

During the collectivist era, the dynamism of master craftsmen in its neighbourhood created favourable conditions for the emergence of a new craft in Dông Ky: that of fine furniture. With strong links to Eastern European markets, they fulfilled orders placed by the State. The State provided them with raw materials, machinery and took care of exporting. The number of artisans grew rapidly.

At the same time, former Dông Ky buffalo traders began to trade in wood, while still others invested part of their capital into buying antique furniture that the Hanoian bourgeoisie, fleeing the communist regime, sold them cheaply. The traders in antique furniture thus initiated a new craft: fine cabinetmaking. They scoured the Delta provinces looking for antique Vietnamese, Chinese, and even French pieces of furniture, then dismantled and copied them. To begin with, they employed the qualified artisans and cabinetmakers from the surrounding craft villages or from other Delta provinces, to whom they paid very good salaries. By frequenting them, the children and apprentices of the Dông Ky bosses learnt this art. One of the first founders of this craft willingly admits "the people of Dông Ky stole the craft of other villages".

In 2006, more than 26,800 artisans and labourers were working in the cluster driven by Dông Ky, either in the workshops and stores of this village, or at home in their own workshops. 40% of the workforce comes from the village, without counting the many workshops that operate by subcontracting for it and that are set up in the neighbouring villages, space being at a premium (Fanchette and Nguyên Xuân Hoan 2009).

Fine cabinetmaking is very largely a manual activity, only a tiny part of the production process is mechanized. It employs a workforce with diversified levels of qualification: sculptors, engravers, mother-of-pearl inlayers, cabinetmakers, and a host of lowly qualified workers that carry out the finishing work (sanding down, varnishing, assembly, etc.). The various stages of production are shared out among several villages and kinds of artisans. Most of the businesses subcontract the most ornate and carved parts of pieces of furniture to the cluster's specialized artisans. The easiest production stages are performed by families new to the craft. They carry out a tiny part of the production process (the legs or the armrests of armchairs, the backs of chairs, etc.). They must also subcontract the most ornate parts of furniture to specialized artisans from the cluster's villages.

Dông Ky businesses employ specialized artisans in their workshops who come from distant villages: qualified cabinetmakers and mother-of-pearl inlayers. Since the beginning of this century, 500 mother-of-pearl inlayers have lived in Dông Ky renting workshops to carry out subcontracted orders.

Businesses in the mother-village employ workers (mostly women) that they train through apprenticeship in their workshops to assemble the subcontracted pieces, carry out finishing work (sanding and varnishing), cutting wood or fulfilling small orders.

They also buy ready-made furniture in the villages of Huong Mac and Phu Khê to make up their orders, capitalizing on their legal and financial capabilities and their commercial contacts for export.

The internationalization of Dông Ky came about thanks to the emergence of "export agents", controlling the cluster's relations with the outside world. For the most part, these people were heads of family businesses that have become officially declared, or were integrated into networks, for example military (during the two wars, many soldiers acquired knowhow, travelling all over Vietnam and neighbouring countries), political (the former co-operatives networks from the collectivist era are still very active and were crucial when the State privatized trade) or commercial and familial. They have succeeded in penetrating the entire sub-regional networks of trade in wood and fine furniture.

5. Conclusion

Since their inception, the craft villages of the Red River Delta have developed on the basis of their contacts with the "District of the 36 streets" in Hà Nôi and with a tight network of rural markets. In contact with the regions on higher ground producing plant raw materials (bamboo, rattan, wood, lacquer and so on) through a dense river network and opening out onto the outside world through the traders of Hà Nôi, they managed to prosper because they enjoyed diversified markets, particularly the luxury one for the Imperial Court, and an abundant workforce. Their links with China, their former master, galvanized exchanges of technical knowhow and enabled innovation in certain fields of fine crafts.

The artisans of craft villages managed to survive the upheavals of the country's economic history and to adapt to very varied markets, increasing and diversifying their production. Despite great weaknesses in terms of capital and perfunctory training, most of these artisans work within informal and specialized familial workshops that can produce items for the international market by subcontracting for officially declared businesses.

The key to this success is their organization into clusters, where the division of labour and specialization enable them to limit competition between villages and to create a very flexible production system. At the top of the pyramid, those who place orders, mostly organized into officially declared companies or businesses, subcontract to a host of informal small family businesses manual or labour-intensive parts of the items to be made.

With the economic liberalization of the 1980s, craft villages encountered several challenges owing to a very rapid growth in production and to a diversification in products to respond to fresh markets. Without training suited to this economic context, the most dynamic artisans mechanized their production, officially declared their businesses and joined trading networks, particularly those for export. Having lost almost all links with the "District of the 36 streets" since the collectivist era, without the support of craft guilds, they are developing with no other protection than membership of a production chain where the risks are shared out among all its members.

But these production chains change rapidly with market constraints, particularly international ones that are very demanding on quality and production deadlines. The difficulty in controlling the production process and obtaining compliance with precise technical specifications from artisans who only make one part of an item is one of the defects of this nevertheless very dynamic and flexible system. In addition, the need to mechanize and standardize certain parts of the process has encouraged a minority of entrepreneurs to integrate the production chain into their workshop and to eradicate this division of labour system by rigidifying it. At the same time, mechanization of some stages tends to lengthen the commodity chain because some artisans are specializing in the use of a machine from which they seek to get a maximum return. These two conflicting processes threaten the future of the cluster system and by the same token that of craft villages.

In the first case, these businesses seek to resemble modern capitalist or State enterprises that are protected by social legislation and enjoy fiscal advantages and access to land. To produce more cheaply than such enterprises is their credo, which incurs the risk of no longer enjoying the flexibility of the cluster system. The danger is that the economic crisis may hit them hard and drag them under, as was the case for several officially declared businesses in the wickerwork sector in 2008.

In the second case, lengthening the commodity chain renders the value of relations between members more fragile, less confident and more complex, and limits the possibility of quality labelling of products with poor traceability. Vietnam's entry into the WTO will doubtless reinforce the vertical integration process of businesses and limit subcontracting.

It might have been expected that the growth in production and the official recognition of many businesses might have had some influence on the types of co-operation between businesses and changed the nature of the social capital within these clusters. Professional associations, service co-operatives, support from local authorities and technical services to improve the conditions of production are examples of official bodies that, in Western clusters, strengthen horizontal links between businesses of various sizes and makes the cluster gel together. It is apparently still too early to hope for something similar for Vietnam!